5 STEPS TO A 5

500
AP Physics 2 Questions
to know by test day

Also in the McGraw Hill *500 Questions* Series
500 ACT English and Reading Questions to know by test day
500 ACT Math Questions to know by test day
500 ACT Science Questions to know by test day
500 GRE Math Questions to know by test day
500 HESI A2 Questions to know by test day
500 SAT Reading, Writing, and Language Questions to know by test day
500 SAT Math Questions to know by test day
5 Steps to a 5: 500 AP Biology Questions to know by test day
5 Steps to a 5: 500 AP Calculus AB/BC Questions to know by test day
5 Steps to a 5: 500 AP Chemistry Questions to know by test day
5 Steps to a 5: 500 AP English Language Questions to know by test day
5 Steps to a 5: 500 AP English Literature Questions to know by test day
5 Steps to a 5: 500 AP European History Questions to know by test day
5 Steps to a 5: 500 AP Human Geography Questions to know by test day
5 Steps to a 5: 500 AP Macroeconomics Questions to know by test day
5 Steps to a 5: 500 AP Microeconomics Questions to know by test day
5 Steps to a 5: 500 AP Physics 1 Questions to know by test day
5 Steps to a 5: 500 AP Physics 2 Questions to know by test day
5 Steps to a 5: 500 AP Physics C Questions to know by test day
5 Steps to a 5: 500 AP Psychology Questions to know by test day
5 Steps to a 5: 500 AP Statistics Questions to know by test day
5 Steps to a 5: 500 AP U.S. Government & Politics Questions to know by test day
5 Steps to a 5: 500 AP U.S. History Questions to know by test day
5 Steps to a 5: 500 AP World History Questions to know by test day

5 STEPS TO A 5

500 AP Physics 2 Questions
to know by test day
Second Edition

Christopher Bruhn

New York Chicago San Francisco Athens London Madrid
Mexico City Milan New Delhi Singapore Sydney Toronto

Copyright © 2022, 2017 by McGraw Hill. All rights reserved. Printed in the United States of America. Except as permitted under the United States Copyright Act of 1976, no part of this publication may be reproduced or distributed in any form or by any means, or stored in a database or retrieval system, without the prior written permission of the publisher.

1 2 3 4 5 6 7 8 9   LCR   26 25 24 23 22

ISBN   978-1-264-27500-7
MHID   1-264-27500-5

e-ISBN   978-1-264-27501-4
e-MHID   1-264-27501-3

McGraw Hill, McGraw Hill logo, 5 Steps to a 5, and related trade dress are trademarks or registered trademarks of McGraw Hill and/or its affiliates in the United States and other countries and may not be used without written permission. All other trademarks are the property of their respective owners. McGraw Hill is not associated with any product or vendor mentioned in this book.

AP, Advanced Placement Program, and College Board are registered trademarks of the College Board, which was not involved in the production of, and does not endorse, this product.

McGraw Hill products are available at special quantity discounts to use as premiums and sales promotions or for use in corporate training programs. To contact a representative, please visit the Contact Us pages at www.mhprofessional.com.

McGraw Hill is committed to making our products accessible to all learners. To learn more about the available support and accommodations we offer, please contact us at accessibility@mheducation.com. We also participate in the Access Text Network (www.accesstext.org), and ATN members may submit requests through ATN.

DEDICATION

This book is dedicated to all the AP Physics 2 teachers who toil alone, because they are the only one of their kind at their school. They have no one else to work with, share with, or talk to about their subject. Yet, they love the incredible beauty and wonder of physics and how it all fits together. I hope this book helps you and your students.

CONTENTS

Introduction viii
About the Author xi
Acknowledgments xii

Diagnostic Quiz 1
Getting Started: The Diagnostic Quiz 3
Diagnostic Quiz Questions 5
Diagnostic Quiz Answer Explanations 21

Chapter 1 Fluids 29
Questions 1–57

Chapter 2 Thermodynamics and Gases 53
Questions 58–118

Chapter 3 Electric Force, Field, and Potential 85
Questions 119–196

Chapter 4 Electric Circuits 119
Questions 197–265

Chapter 5 Magnetism and Electromagnetic Induction 147
Questions 266–331

Chapter 6 Geometric and Physical Optics 171
Questions 332–420

Chapter 7 Quantum, Atomic, and Nuclear Physics 205
Questions 421–500

Answers 223

INTRODUCTION

How to use this book and words of wisdom for **STUDENTS:**

The AP Physics 2 exam is not for the faint of heart.... There are deep conceptual problems, challenging symbolic mathematical questions, laboratory design scenarios, and thought-provoking questions that require you to justify your answers in scientific detail. These are all skills that can be mastered with practice. You want to get good at AP Physics 2 and pass the AP exam? Just like anything else, it takes preparation and persistence. This book is designed to give the extra training necessary to excel on the exam. These are the same questions that I use with my own students to help them defeat the AP Physics 2 exam.

First, use this book to review each content section as you cover it in your AP Physics 2 class. This will help to deepen your knowledge of each subject area with added practice AND improve your class grade at the same time. Second, use this book as a review of the course content as you prepare for the exam. Just remember that understanding AP Physics takes time and effort. The universe was not created in a day and it certainly cannot be understood in a one-night cram session. You simply cannot learn physics overnight, so don't start studying May 1st. Give your brain some time to digest how it all fits together.

Two very important points:

1. Don't look at the answers until you have truly attempted to answer the question. I know that you come from the Google Generation where all the answers are at your fingertips, but brain research tells us that looking at the answer doesn't help you learn nearly as well as struggling to answer the question. **Work through the questions on your own or in a group before you look at the answer.**
2. Most of the AP Physics questions ask you to show your work and write out your justifications and explanations to your answer. This is very important. Every year many students fail the AP Physics exam because they don't know how to organize their thoughts and put it down on paper in "physics speak." **Write out your explanations!** I've given you examples of what the AP readers are looking for in the ANSWER KEY.

Here are some inspirational quotes to help you get started on your climb to enlightenment:

> "The journey of a thousand miles begins with one step." Lao Tzu
>
> "When eating an elephant take one bite at a time." Creighton Abrams
>
> "Get off your butt and start studying! This physics isn't going to learn itself." C. Bruhn

If you work your way through this book, I promise you will improve and do better on the exam. *Seriously ... I promise!*

How to use this book and words of wisdom for **TEACHERS:**

This book is designed for you!

This started as a project to create 500 AP Physics 2 questions for students to use for review, but it has transformed into much more. Here is why:

- I quickly realized that 500 questions was not enough to adequately cover all the AP Physics 2 content. So, many of the questions you find in this book have grown into comprehensive multipart affairs that attack critical concepts from different angles.
- Simply writing more AP-style questions was going to be counterproductive. There are already good resources including the *5 Steps to a 5 AP Physics 2* book that I hope you already have.
- There are very few resources with questions that actually help students build the skill they need to excel on the AP Physics 2 exam. Textbook questions seem to be either too numerically intensive or the conceptual questions are not up to AP level of difficulty. So, teachers get stuck trying to use old AP Physics 2 questions, but there never seems to be enough of those. In addition, many of the released AP exam questions are really hard for students to grasp without the groundwork leading up to them.

With this realization, I wrote this book for myself and other AP Physics 2 teachers. The questions in this book are an integral part of my own lesson plans. Everything on the pages that follow is material I use with my own students to build their skill and prepare them for the exam in May. For each content area, there are skill-building questions followed by AP-style questions. There are enough AP-style questions to create two complete multiple-choice practice exams and over four complete free-response practice exams.

Every single College Board AP Physics 2 learning objective (LO) is covered. Most LOs are addressed in multiple ways from different angles. In most of the questions I have asked students to WRITE OUT THEIR EXPLANATION. This cannot be emphasized enough. Many students fail the exam simply because they don't have enough practice writing in "physics speak" and organizing their thoughts logically. Please make your students write in class. Make them logically defend their thoughts in writing. Then, have them read and critique what other students have written. This book has paragraph-length response questions, laboratory design questions, multiple correct questions, qualitative-quantitative transition questions, ranking tasks, etc. In short, it has everything you need to help your students improve on the exam.

I hope you enjoy the book and that it serves you and your students well!

ABOUT THE AUTHOR

Chris Bruhn began his career as an aerospace engineer before becoming a physics teacher. Since becoming an educator, he has taught all varieties of AP Physics in all types of schools and has won several educational awards. Chris is an educational trainer. He likes to create and share curriculum and educational resources as well as lead AP summer institutes and study sessions for teachers and students around the country. Outside of teaching, Chris likes building things and tearing things apart! He enjoys sports, painting, travel, watching superhero movies with his kids, and generally having fun. And now he has written this book!

ACKNOWLEDGMENTS

I'd like to thank the great Mite Munce and spell check without which this would not be possible.

Thank you to Kern for field-testing these questions with his own students and checking me for errors because everybody makes mistakes.

But most of all thanks to my family who put up with me sequestered in the office for weeks on end writing this ridiculous beast. They are the best even when I am the worst.

Diagnostic Quiz

GETTING STARTED: THE DIAGNOSTIC QUIZ

The following is a 20-question quiz to gauge where you presently stand in your knowledge of the AP Physics 2 content. It is important that you take the quiz by yourself and not use outside resources like the internet or your friends. Many times, students fool themselves into thinking that they know more than they do because they are looking at the key or they are using a resource. Remember that the AP test is taken alone. By yourself. Just you and your brain. It's intimidating for students who are used to easy access to the internet. I want you to be prepared and confident when you walk into the AP test. Knowledge is power. You can do it.

The quiz covers all the content areas. The questions are similar to what you will see on the AP exam—some multiple choice, both select-one and select-two questions, and some free response. Give yourself about an hour to complete the quiz.

It's time to see what you already know and what you need to gain a better understanding of. So take the quiz and plan your path to mastery of AP Physics 2.

Good luck!

DIAGNOSTIC QUIZ QUESTIONS

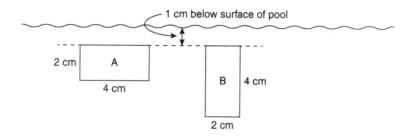

1. Two identical blocks float 1 cm below the surface of a swimming pool. One of the blocks is floating horizontally, and the other is floating vertically, as seen in the figure. Which of the following correctly describes the difference in water pressure between the top and bottom surfaces of the two blocks and the buoyancy force experienced on the two blocks by the water?

	Difference in water pressure between the top and bottom of the block	Buoyancy force on the block from the water
(A)	$\Delta P_A = \Delta P_B$	$F_{buoy\,A} = F_{buoy\,B}$
(B)	$\Delta P_A = \Delta P_B$	$F_{buoy\,A} < F_{buoy\,B}$
(C)	$\Delta P_A < \Delta P_B$	$F_{buoy\,A} = F_{buoy\,B}$
(D)	$\Delta P_A < \Delta P_B$	$F_{buoy\,A} < F_{buoy\,B}$

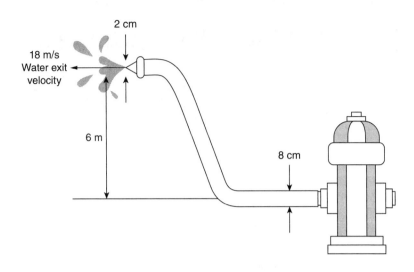

2. Firefighters are using a hose with a 2-cm-diameter exit nozzle connected to a hydrant with an 8-cm-diameter opening to combat a fire on the second floor of a building 6 m above the hydrant, as shown in the figure. What pressure must be supplied at the hydrant to produce an exit velocity of 18 m/s? (Assume that the density of water is 1,000 kg/m^3 and that the exit pressure is 1×10^5 Pa.)

(A) 1.7×10^5 Pa
(B) 2.0×10^5 Pa
(C) 2.6×10^5 Pa
(D) 3.2×10^5 Pa

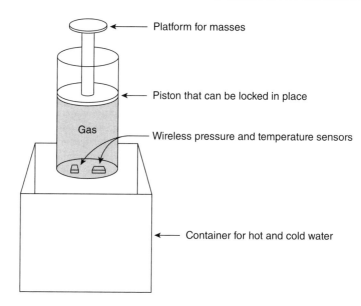

3. Students have the following test apparatus:
   - A cylinder that contains a sample of ideal gas. The cylinder has a piston that moves freely or can be locked in place.
   - Wireless temperature and pressure sensors are installed inside the gas cylinder.
   - The piston has a platform where masses can be added.
   - The piston can be lowered into a container containing hot or cold water.

   The gas sample begins at room temperature and atmospheric pressure. The students take the gas sample through a cycle ABCA.

   - In process AB, the gas is expanded to twice its original volume while constant pressure is maintained in the gas.
   - In process BC, the gas is brought back to its original volume isothermally.
   - In process CA, the gas returns to its original state without any work being done on the gas.

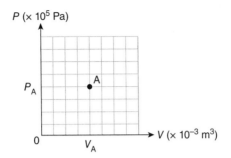

(A) On the axis, sketch a graph of the pressure P as a function of volume V for the process ABCA. The starting pressure and volume P_A and V_A are given. Label states B and C on the diagram. Also label the values for the pressure and volume for states B and C on the axes in terms of P_A and V_A.
(B) For process AB, the gas cylinder is lowered into a water bath. Does the water bath contain hot or cold water? Support your claim using physics principles.
(C) For process BC, is energy being added to the gas or removed from the gas? Support your claim using physics principles.
(D) For process CA, is the entropy of the gas increasing, decreasing, or remaining the same? Support your claim using physics principles.
(E) For which processes, if any, was weight added to the platform? Support your claim using physics principles.

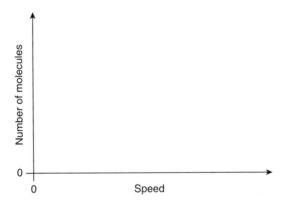

(F) On the axis, sketch and label the distribution of the gas molecule speed when the gas is in state A, B, and C. Support your sketch using physics principles.

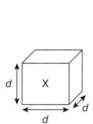
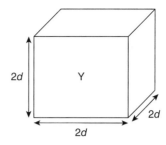

4. Two cubes of copper sit on a lab table side by side, as shown in the figure. Cube X has a temperature of 800 K. Cube Y has a temperature of 400 K. Which of the following correctly describes what happens when cube X is set on top of cube Y? *Select two answers.*
   (A) Heat flows from cube Y to cube X because cube Y has more total internal energy than cube X.
   (B) Heat flows from cube X to cube Y because, on average, the atoms in cube X are moving faster.
   (C) Heat flows between the cubes until both cubes have the same total internal energy.
   (D) Heat flows between the cubes until the atoms in both cubes have the same average kinetic energy.

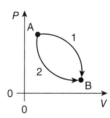

5. A gas can be taken from an initial state A to a final state B along two different processes, as shown in the *PV* diagram. Which of the following is correct concerning the magnitude of the change in internal energy and the magnitude of the work in the two processes?
   (A) $\Delta U_1 = \Delta U_2$ and $W_1 = W_2$
   (B) $\Delta U_1 = \Delta U_2$ and $W_1 > W_2$
   (C) $\Delta U_1 > \Delta U_2$ and $W_1 = W_2$
   (D) $\Delta U_1 > \Delta U_2$ and $W_1 > W_2$

6. Two identical neutral metal spheres are touching, as shown in the figure. Which of the following locations of a positively charged insulating rod will create the largest positive charge in the sphere on the right?

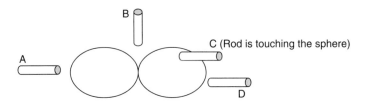

7. A student brings a negatively charged rod near an aluminum sphere but does not touch the rod to the sphere. He grounds the sphere and then removes the ground. Which of the following correctly describes the force between the rod and sphere before and after the sphere is grounded?

	Before Grounding	After Grounding
(A)	attraction	attraction
(B)	attraction	repulsion
(C)	no force	attraction
(D)	no force	no force

8. Three small droplets of oil with a density of ρ are situated between two parallel metal plates, as shown in the figure. The bottom plate is charged positive, and the top plate is charged negative. All the particles begin at rest. As time passes, particle 1 accelerates downward, particle 2 remains stationary, and particle 3 accelerates upward, as shown. Which of the following statements is consistent with these observations?

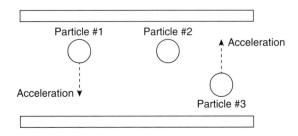

(A) Particle 1 must be negatively charged.
(B) Particle 2 must have no net charge.
(C) Particle 2 has a mass that is too small to affect its motion.
(D) Particle 3 must be positively charged.

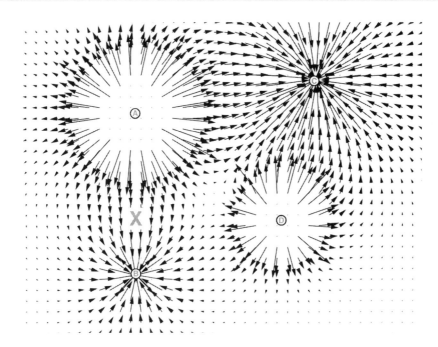

9. The figure above represents the electric field in the region around four small objects A, B, C, and D.

    (A) Indicate the sign of each charge by checking the appropriate box below.

	Positive	Negative	Neutral
Charge A			
Charge B			
Charge C			
Charge D			

    Describe the characteristics of the field that indicate the sign of the objects.

    (B) Rank the objects from most positive to most negative. Describe the characteristics of the field that allow you to correctly make this ranking.

    (C) An electron is placed at location X and released. In what direction will the electron move immediately after being released?

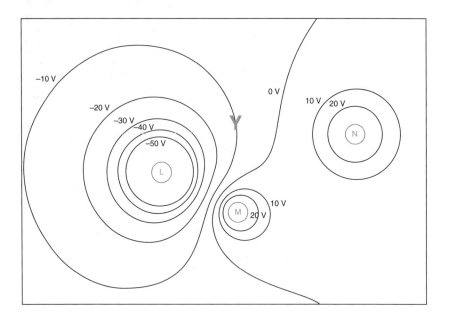

The figure above depicts equipotential lines around three small objects L, M, and N.

(D) Indicate the sign of each object by checking the appropriate box below.

	Positive	Negative	Neutral
Charge L			
Charge M			
Charge N			

(E) Does it require work to move a 2-C charge from the −30-V equipotential line around object L to the 20-V equipotential line surrounding object N? If so, calculate the amount of work required. If not, explain why no work is required.

(F) In what direction will an electric field vector point when drawn at location Y? Explain your response.

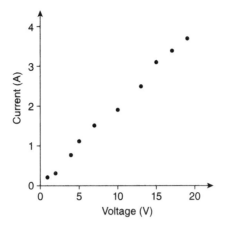

10. A kitchen roaster is connected to a variable power supply. The voltage difference across and current through the roaster are measured for various settings of the power supply. The figure above shows the graph of the data. The resistance of the toaster
    (A) varies up and down with voltage.
    (B) increases linearly with input voltage.
    (C) is constant at 0.2 Ω.
    (D) is constant at 5.0 Ω.

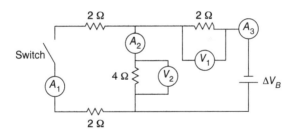

11. The circuit shown above has a battery of negligible internal resistance, resistors, and a switch. There are voltmeters, which measure the potential differences V_1 and V_2, and ammeters A_1, A_2, and A_3, which measure the currents I_1, I_2, and I_3. The switch is initially in the closed position. With the switch still closed, which of the following relationships are true? *Select two answers.*
    (A) $I_1 + I_2 - I_3 = 0$
    (B) $\Delta V_B - V_1 - V_2 = 0$
    (C) $V_1 > V_2$
    (D) $I_2 = I_3$

14 › 500 AP Physics 2 Questions to know by test day

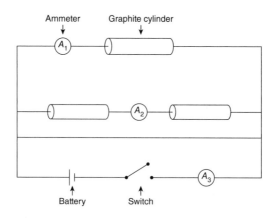

12. A physics student assembles a circuit like that shown in the figure using three identical graphite cylinders as resistors. The graphite resistors have resistivity r, length L, and diameter D. There are three ammeters A_1, A_2, and A_3 in the circuit that measure currents I_1, I_2, and I_3, respectively. After setting up the circuit, the student closes the switch and measures the currents.

   (A) Rank the currents I_1, I_2, and I_3 from greatest to least. Explain your ranking using physics principles.
   (B) It is possible to make the current the same in two of the ammeters. In which two ammeters can this be accomplished? This is done by modifying one or more of the circuit elements without removing them. Describe how one or more or the circuit elements could be modified to accomplish this. Explain your reasoning using physics principles.

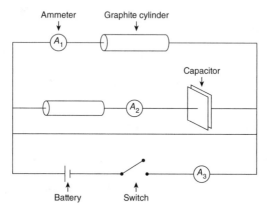

A new circuit is constructed with two identical graphite cylinders and one capacitor, as shown in the figure. There are three ammeters A_1, A_2, and A_3 in the circuit that measure currents I_1, I_2, and I_3, respectively. The switch is in the open position, and the capacitor is uncharged.

   (C) Describe the currents measured by each ammeter from the time the switch is closed until a long time has passed.

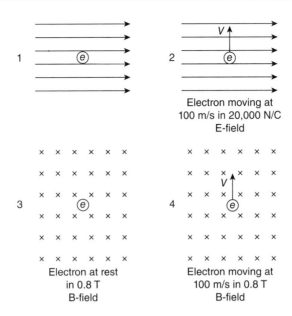

13. The figure above shows two electric fields and two magnetic fields with an electron imbedded in each field. In cases 1 and 3, the electron begins at rest. In cases 2 and 4, the electron is moving with a velocity of 70,000 m/s.

(A) In case 1, calculate the force on the electron. Indicate the direction of the force with an arrow. Indicate the direction of the electron's path using a dashed line, and describe the shape of the path.

(B) In case 2, calculate the force on the electron. Indicate the direction of the force with an arrow. Indicate the direction of the electron's path using a dashed line, and describe the shape of the path.

(C) In case 3, calculate the force on the electron. Indicate the direction of the force with an arrow. Indicate the direction of the electron's path using a dashed line, and describe the shape of the path.

(D) In case 4, calculate the force on the electron. Indicate the direction of the force with an arrow. Indicate the direction of the electron's path using a dashed line, and describe the shape of the path.

16 › 500 AP Physics 2 Questions to know by test day

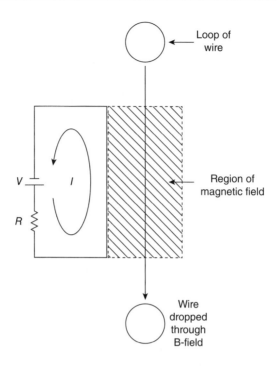

14. A circuit consisting of a battery, a resistor, and wire is connected as shown. The current passing around the circuit as indicated produces a magnetic field in the shaded region of the figure. A loop of wire is held above the magnetic field and dropped. Which of the following correctly indicates the direction of the magnetic field in the shaded region of the figure, and the location where current is produced in the loop of wire as it falls downward?

	Direction of the B-field produced by the circuit	Current is induced in the falling loop of wire
(A)	Into the page	When entering the B-field
(B)	Into the page	When in the middle of the B-field
(C)	Out of the page	When in the middle of the B-field
(D)	Out of the page	When exiting the B-field

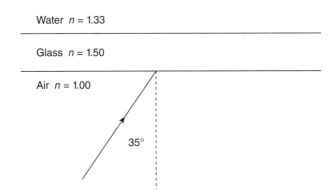

15. Light traveling through air encounters a glass aquarium filled with water. The light is incident on the glass from the front at an angle of 35°.
    (A) At what angle does the light enter the glass?
    (B) At what angle does the light enter the water?
    (C) On the diagram above, sketch the path of the light as it travels from air to water. Include all reflected and refracted rays; label all angles of reflection and refraction.
    (D) Which physical properties of the light change as the light travels from the air into the glass? Explain your reasoning.

16. Two waves are traveling toward each other, as shown in the figure. Which of the following are possible results of the wave patterns as the two waves pass by each other? *Select two answers.*

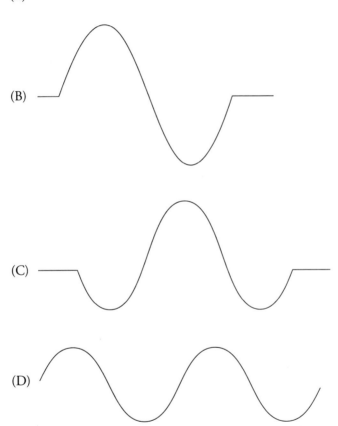

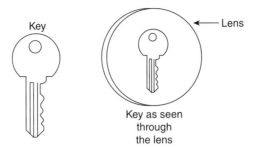

17. A student looks at a key through a lens. When the lens is 10 cm from the key, what the student sees through the lens is shown in the figure. The student estimates that the image is about half the size of the actual key. What is the approximate focal length of the lens being used by the student?
    (A) −10 cm
    (B) −0.1 cm
    (C) 0.3 cm
    (D) 3.0 cm

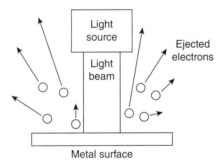

18. A beam of ultraviolet light shines on a metal plate, causing electrons to be ejected from the plate, as shown in the figure. The velocity of the ejected electrons varies from nearly zero to a maximum of 1.6×10^6 m/s. If the brightness of the beam is increased to twice the original amount, what will be the effect on the number of electrons leaving the metal plate and the maximum velocity of the electrons?

	Number of electrons ejected	Maximum velocity of ejected electrons
(A)	Increases	Increases
(B)	Increases	Remains the same
(C)	Remains the same	Increases
(D)	Remains the same	Remains the same

$$^{12}_{6}\text{C} + ^{1}_{1}\text{H} \rightarrow ^{13}_{7}\text{N} + \gamma$$

19. Which of the following expressions correctly relates the masses of the constituent particles involved in the nuclear reaction shown?

    (A) $m_C + m_H - m_N = 0$

    (B) $m_C + m_H - m_N - \dfrac{hf_\gamma}{c^2} < 0$

    (C) $m_C + m_H - m_N - \dfrac{hf_\gamma}{c^2} = 0$

    (D) $m_C + m_H - m_N - \dfrac{hf_\gamma}{c^2} > 0$

20. Which of the following phenomena can be better understood by considering the wave properties of electrons? *Select two answers*

    (A) There are discrete electron energy levels in a hydrogen atom.
    (B) Monochromatic light of various intensities ejects electrons of the same maximum energy from a metal surface.
    (C) A beam of electrons reflected off the surface of a crystal creates a pattern of alternating intensities.
    (D) An x-ray colliding with a stationary electron causes it to move off with a velocity.

DIAGNOSTIC QUIZ ANSWER EXPLANATIONS

1. (Chapter 1: Fluids)

(C) Static fluid pressure is given by the equation

$$P = P_0 + \rho g h$$

The difference in pressure between the top and the bottom of the block will be

$$\Delta P = \rho g h_{bottom} - \rho g h_{top} = \rho g \Delta h$$

Therefore, the pressure differential between the top and bottom will be greatest for block B. The buoyancy force ($F_b = \rho V g$) will be the same for both blocks because they have the same volume while the density of water and acceleration of gravity are the same in both cases.

2. (Chapter 1: Fluids)

(D) Using the conservation of mass/continuity equation, we see that the water must be slower at the hydrant than when exiting the nozzle:

$$A_1 v_1 = A_2 v_2$$

The area of the hose is proportional to the radius squared:

$$\pi r_1^2 v_1 = \pi r_2^2 v_2$$

$$\pi \left(\frac{8 \text{ cm}}{2}\right)^2 v_1 = \pi \left(\frac{2 \text{ cm}}{2}\right)^2 18 \text{ m/s}$$

This gives us a velocity in the hose of 1.125 m/s. Using conservation of energy/Bernoulli's equation and assuming that the exit pressure is atmospheric, we get

$$\left(P + \rho g y + \frac{1}{2}\rho v^2\right)_1 = \left(P + \rho g y + \frac{1}{2}\rho v^2\right)_2$$

$$\left[P + 0 + \frac{1}{2}(1{,}000 \text{ kg/m}^3)(1.125 \text{ m/s})^2\right]_1$$

$$= \left[100{,}000 \text{ Pa} + (1{,}000 \text{ kg/m}^3)(10 \text{ m/s}^2)(6 \text{ m}) + \frac{1}{2}(1{,}000 \text{ kg/m}^3)(18 \text{ m/s})^2\right]_2$$

$$P = 320{,}000 \text{ Pa} = 3.2 \times 10^5 \text{ Pa}$$

3. (Chapter 2: Thermodynamics and Gases)

(A)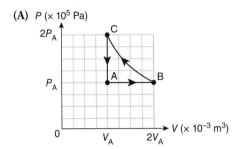

(B) We can see from the PV diagram that the volume doubles while the pressure stays the same. Using the ideal gas law ($PV = nRT$), we can see that the temperature needs to increase. This is accomplished with a hot-water bath.

(C) Using the ideal gas law ($PV = nRT$), we can see that the temperature remains constant because the PV value is the same at states B and C. This means that the internal energy of the gas remains the same, that is, $\Delta U = \frac{3}{2} nR\Delta T$. We also know that the work is positive because the volume change is negative ($W = -P\Delta V$). Using the first law of thermodynamics ($\Delta U = Q + W$), we can deduce that heat Q must be leaving the gas. Therefore, energy is being removed from the gas.

(D) In process CA, no work is done. The temperature and internal energy of the gas must decrease. Using the first law of thermodynamics ($\Delta U = Q + W$), we can deduce that heat must exit the gas. This means that the entropy of the gas is decreasing, that is, $\Delta S = \frac{Q}{T}$.

(E) Adding mass to the platform will compress the gas. Only process BC shows the gas being compressed. Process AB shows expansion, and process CA has a constant volume.

(F) States B and C have the same temperature and therefore will have the same distribution. State A has a lower temperature than states B and C. Therefore, the distribution for state A will have a peak at a lower speed:

$$v_{rms} = \sqrt{\frac{3k_B T}{m}}$$

The distributions represent the same quantity of gas, and the areas under the curves should be the same. Therefore, the peak of the state A distribution should be higher than for states B and C.

4. (Chapter 2: Thermodynamics and Gases)

(B and D) The higher the temperature of an object, the higher the average kinetic energy of the atoms will be. These faster-moving atoms tend to transfer energy to slower-moving atoms when in thermal contact. Heat will transfer from the hotter object to the colder object until both have the same temperature and the atoms have the same average kinetic energy.

5. (Chapter 2: Thermodynamics and Gases)

(B) Both paths start and end at the same point. Therefore, the initial and final temperatures are the same, as are the initial and final thermal energies. Process 1 has a higher average pressure for the same volume change. Another way to think about this is to compare the areas under the curves. Process 1 has more area under its curve than process 2 and therefore a larger magnitude of work. (Note that the work is negative in both cases, but we are asked to compare the magnitude of the work.)

6. (Chapter 3: Electric Force, Field, and Potential)

(A) Rods A, B, and D will each polarize the spheres, drawing negative charges toward themselves and leaving the opposite side positively charged. Thus A will cause the right sphere to be the most positive. Touching the spheres with the insulating rod will cause some of the polarized negative charge from the spheres to flow onto the rod. Because the rod is an insulator, this leaves the spheres with only a small excess positive charge that will be shared between both spheres.

7. (Chapter 3: Electric Force, Field, and Potential)

(A) Before grounding, the negatively charged rod polarized the sphere, causing an attraction. After grounding, the sphere has been charged the opposite sign by the process of induction, and the two will attract.

8. (Chapter 3: Electric Force, Field, and Potential)

(D) All the droplets have mass ($m = \rho V$) and will experience a downward gravitational force. Particle 1 could be uncharged and simply fall due to the force of gravity. Particle 2 must have an electric force to cancel the gravity force. Particle 3 must be positive to receive an electric force upward larger than the force of gravity downward.

9. (Chapter 3: Electric Force, Field, and Potential)

(A)

	Positive	Negative	Neutral
Charge A	X		
Charge B		X	
Charge C		X	
Charge D	X		

Electric field vectors point away from the positive charges and inward toward negative charge.

(B) (Most positive) A > D > B > C (most negative). The electric field vectors that point away indicate positive charge, and the vectors that point inward are negative charge. To gauge the size of the charge, simply look at the electric field vectors at a given distance from the object. The field vectors that are the longest for a given distance will indicate the strongest charge.

(C) Upward toward charge A. The electron is negative and will receive a force in the opposite direction of the electric field.

(D)

	Positive	Negative	Neutral
Charge L		X	
Charge M	X		
Charge N	X		

The electric potential becomes more positive the nearer we get to M and N, indicating that they are positive. The electric potential becomes more negative the nearer we get to L indicating it is negative.

(E) A positive charge will naturally move from more positive potential to more negative potential. So, yes, work is required to move the positive 2-C charge to a more positive location:

$$W = \Delta U_E = q\Delta V = q(V_f - V_i) = (2 \text{ C})((20 \text{ V}) - (-30 \text{ V})) = 100 \text{ J}$$

(F) The electric field vector will point directly left. Electric field vectors are always perpendicular to equipotential lines and point toward decreasing voltage.

10. (Chapter 4: Electric Circuits)

(D) There is some data scatter, but the data appear to be linear with a slope of about 1/5. Current, as a function of voltage, is given by $I = \dfrac{\Delta V}{R} = \dfrac{1}{R}\Delta V$. Thus the slope of the line should equal 1/R. Therefore, the resistance is approximately 5 Ω.

11. (Chapter 4: Electric Circuits)

(A and B) Applying Kirchhoff's junction rule to the node above the A_2 ammeter, we get

$$\Sigma I_{junction} = I_1 + I_2 - I_3 = 0$$

Applying Kirchhoff's current rule to the loop on the right of the circuit containing the battery, we get

$$\Sigma \Delta V_{loop} = \Delta V_B - \Delta V_1 - \Delta V_2 = 0$$

$I_2 < I_3$: Ammeter 3 is in the main circuit line and carries the total current. This current splits, with half going through ammeter 1 and 2.

$V_1 = V_2$: If you work out the math, you will see that these voltages are the same.

12. (Chapter 4: Electric Circuits)

(A) $I_3 > I_1 > I_2$. Using Kirchhoff's current rule, we can see that current 3 is the addition of currents 1 and 2 and is the greatest. Ammeters 1 and 2 are placed in parallel with the same voltage. The current in ammeter 1 is greater than that in ammeter 2 because the resistance is twice as large in the ammeter 2 line because of the two resistors in series.

(B) It is possible to make the currents the same in ammeters 1 and 2. This can be accomplished by making the resistance the same in both parallel lines by modifying the graphite resistors. To do this, we can do one of the following:

1. Modify the top graphite cylinder to have twice the resistance. Make the top resistor twice as long or reduce the diameter to $D/\sqrt{2}$.
2. Modify the bottom two graphite cylinders to have half the resistance. Make the lower resistors half as long or increase the diameter of the cylinder to $\sqrt{2}D$.

(C) When the capacitor is uncharged, it behaves like a wire and offers no resistance to current flow. Once the capacitor is fully charged, no current passes through it, and it behaves like an open switch or a disconnect in the circuit.

At time equals zero, the circuit behaves like a parallel circuit with one resistor in each line. As time goes by, the line with the capacitor has no current, and the circuit behaves like a series circuit with the current going around the outside loop, and no current passing through the center line. Therefore, the total resistance of the circuit doubles over time, and the current in ammeter 3 decreases with time.

The current in ammeter 2 decreases to zero as the capacitor charges.

The current in ammeter 3 does not change because the voltage supplied to the top resistor does not change.

13. (Chapter 3: Electric Force, Field, and Potential and Chapter 5: Magnetism and Electromagnetic Induction) See figure:

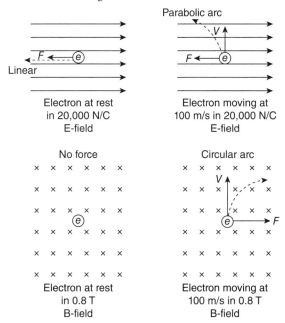

(A) The electric force $F_E = Eq = (20{,}000 \text{ N/C})(1.6 \times 10^{-19} \text{ C}) = 3.2 \times 10^{-15} \text{ N}$
(B) Same as in case (A): $F_E = 3.2 \times 10^{-15} \text{ N}$
(C) No magnetic force
(D) The magnetic force $F_M = qvB = (1.6 \times 10^{-19} \text{ C})(70{,}000 \text{ m/s})(0.8 \text{ T}) = 9.0 \times 10^{-15} \text{ N}$

14. (Chapter 5: Magnetism and Electromagnetic Induction)

(A) On the right side of the circuit, the current is moving upward. Using the right-hand rule for currents, we can see that the B-field in the shaded region is into the page. Current is induced in the falling loop of wire when there is a change in magnetic flux. This occurs when the loop is entering or exiting the B-field.

15. (Chapter 6: Geometric and Physical Optics)

(A) Use Snell's law: $n_1 \sin \theta_1 = n_2 \sin \theta_2$. This becomes $1.0 \sin 35° = 1.5 \sin \theta_2$. Solve for θ_2 to get 22°.

(B) Use Snell's law again. This time, the angle of incidence on the water is equal to the angle of refraction in the glass, or 22°. The angle of refraction in water is 25°. This makes sense because light should bend away from normal when entering the water because water has a smaller index of refraction than glass.

(C) Important points:

- Light both refracts *and reflects* at both surfaces. You must show the reflection, with the angle of incidence equal to the angle of reflection.
- We know you don't have a protractor, so the angles don't have to be perfect. But the light must bend toward normal when entering glass and away from normal when entering water. If you have trouble drawing this on the AP exam, just explain your drawing with a quick note to clarify it for the exam reader. This helps you and the reader!

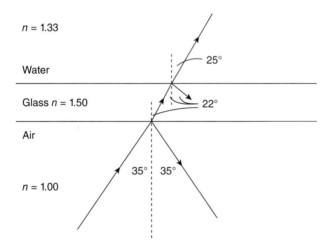

(D) When light enters a new medium, the frequency always stays the same. Because glass has a higher index of refraction than air, the velocity of the light will decrease when entering the glass: $n = c/v$. The wavelength of the light also decreases when entering the glass: $c = f\lambda$.

16. (Chapter 6: Geometric and Physical Optics)

(A and C) The waves will have complete destructive interference when exactly overlapped. When the waves have half passed each other, answer choice **(C)** will be seen.

17. (Chapter 6: Geometric and Physical Optics)

(A) First off, the student is seeing a smaller, upright, virtual image. This means that this has to be a diverging lens with a negative focal length and a negative image distance.

Using the magnification equation

$$M = \frac{s_i}{s_o}, \frac{1}{2} = -\frac{s_i}{10 \text{ cm}}$$ gives an image distance of -5 cm.

Using the lens equation

$$\frac{1}{f} = \frac{1}{s_i} + \frac{1}{s_o} = \frac{1}{-5 \text{ cm}} + \frac{1}{10 \text{ cm}},$$ the focal length equals -10 cm.

18. (Chapter 7: Quantum, Atomic, and Nuclear Physics)

(B) Increasing the brightness of the light increases only the number of photons, not the energy of the individual photons. Thus the number of ejected electrons goes up, but their maximum energy and velocity will still be the same.

19. (Chapter 7: Quantum, Atomic, and Nuclear Physics)

(C) Mass/energy is conserved in nuclear reactions. The mass of the carbon + mass of the hydrogen = mass of the nitrogen + mass equivalent of the gamma ray $m_\gamma = \frac{E_\gamma}{c^2} = \frac{hf}{c^2}$.

20. (Chapter 7: Quantum, Atomic, and Nuclear Physics)

(A and C) The discrete electron energy levels in hydrogen can be understood as being orbits of constructive wave interference locations for the electron. The alternating intensities seen in diffraction patterns are evidence of the wave nature of electrons reflecting off the atoms in the crystal. Both of the other choices are examples of the particle nature of electromagnetic waves.

CHAPTER 1

Fluids

Skill-Building Questions

1. Cube 1 has a density of ρ, volume of V, a mass of m, and a side dimension of x. Cube 2 is made of the same material and has a side dimension of $2x$. What are the density, volume, and mass of cube 2 in terms of ρ, V, and m?

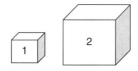

2. Your physics teacher instructs you to determine the density of a rectangular block of unknown material. You are only allowed to use a ruler and a spring scale that measures newtons of force.

    (A) Outline the experimental procedure you would use to gather the necessary data. Indicate the measurements to be taken and how each measurement will be used to obtain the data. Make sure your outline contains sufficient detail so that another student could follow your procedure and duplicate your results.

    (B) Next, your teacher asks you to find the density of an irregularly shaped rock. What additional equipment might you use to determine an accurate density of the rock, and how would you use it?

3. Your teacher has asked you to determine the density of a liquid. You place a graduated cylinder on a balance and measure the mass at different volumes. The data is displayed in the table.

Volume ($\times 10^{-6}$ m^3)	Mass ($\times 10^{-3}$ kg)
50	145
100	161
150	223
200	266
300	334

(A) Graph the data appropriately on the grid provided.

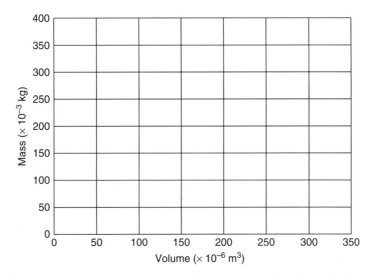

(B) Why does the best fit line not go through the origin? What is the significance of the y-intercept?
(C) Use the graph to determine the density of the fluid in kg/m^3. Show your work.
(D) How might you improve the accuracy of the lab?

4. A block of wood floats in water with one-third of its volume submerged. What does this tell you about the wood?

5. Explain the physical mechanism by which water exerts pressure on the side walls of a glass in which it is contained. Make sure your answer addresses the microscopic scale of matter.

6. The figure shows four cylinders of various diameters filled to different heights with the same fluid. A hole in the side of each cylinder is plugged by a stopper; all of the stoppers are identical. All of the cylinders are open at the top.

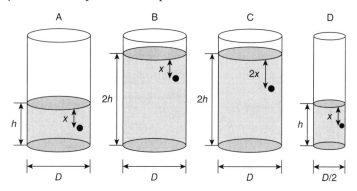

(A) Rank the pressure on the bottom of each cylinder. Justify your claim with an equation.
(B) Rank the force from the fluid on the bottom of each cylinder. Justify your claim with an equation.
(C) Rank the pressure on each stopper. Justify your claim with an equation.

7. Explain Pascal's principle.

Questions 8–10

The figure shows a pump fitted with a piston and a handle. The pump is connected by tubing to three cylindrical containers. The area of each cylinder is given in terms of A (the area of the pump). The heights of the pump and cylinders are given in terms of y. The pump and all the cylinders are completely filled with oil and contain no air pockets. The handle of the pump is pushed down with a force of F.

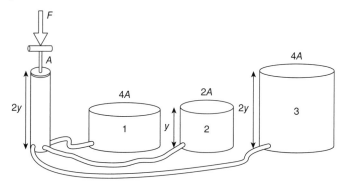

8. What will be the force pushing upward on the top of cylinder 3? Show your work in symbolic form.

9. Explain why the pressure at the top of cylinder 1 will be greater than that at the top of cylinder 3.

10. Which cylinders have the same pressure at the top? Explain.

Questions 11–13

A soda bottle is filled partway with water. A tack is poked into the side to create a hole. The cap is removed from the bottle.

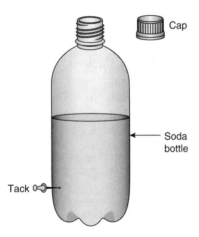

11. When the tack is removed, water flows out. As the water level descends, the velocity of the water leaving the hole decreases. Explain why this happens in terms of pressure and energy.

12. The bottle is refilled to its original level, and the tack is used to plug up the hole. The cap is tightened, and the tack is again removed. Water comes out of the hole but quickly stops while there is still a large quantity of water above the tack hole. Explain why this occurs.

13. This time, the bottle has two tack holes, one above the other, as shown in the figure below. The bottle is filled with water, the cap is tightened, and the bottom tack is removed. Water comes out of the bottom hole but quickly stops while there is still a large quantity of water above both tack holes. What will happen when the top tack is removed? Will water exit either hole? Explain your reasoning.

Questions 14 and 15

A very tall cylinder with a movable piston is placed in a lake. The piston is pulled up, drawing lake water in and upward, as shown in the figure.

14. In a clear, coherent, paragraph-length response, explain why the lake water moves up the cylinder when the piston is pulled upward. Include an explanation of what happens to the air gap above the water in the piston.

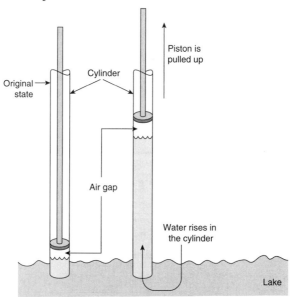

15. Is there a limit to how high water can be drawn up the piston? Write an algebraic expression that supports your answer.

16. Explain how a barometer works.

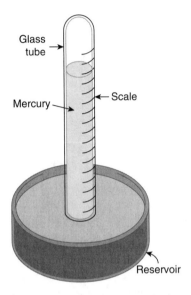

Questions 17 and 18

The figure shows a gas trapped in a spherical container connected to a tube filled with oil that has a density of 930 kg/m³. The top end of the tube is open to the atmosphere, which has a pressure of 100 kPa.

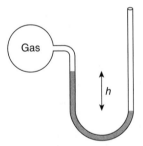

17. Is the pressure of the trapped gas higher or lower than that of the atmosphere? Explain.

18. The oil is higher on one side by the distance $h = 0.22$ m. Calculate the pressure of the trapped gas.

19. Explain the physics mechanism that results in the buoyancy force.

20. Derive an equation to show that the buoyancy force is equal to the weight of water that is displaced by a floating object.

21. The force on the bottom of a swimming pool must increase when a person is floating in the pool. Explain this behavior
    (A) using Newton's third law.
    (B) by referencing the water level in the pool.

22. Explain why objects with a density greater than that of water will sink. Prove your explanation with an equation.

23. If objects that are denser than water sink, how can a steel ship float?

24. A beaker of water rests on an electronic balance, as shown in the figure. A mass suspended from a spring scale is lowered into the water. Location 1 is above the water. Location 2 is just below the surface of the water. Location 3 is just above the bottom of the beaker.

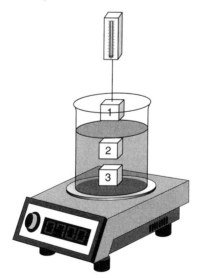

   (A) Do the spring scale and balance readings increase, remain the same, or decrease when the mass is lowered from location 1 to location 2? Explain.
   (B) Do the spring scale and balance readings increase, remain the same, or decrease when the mass is lowered from location 2 to location 3? Explain.

25. You and a friend are taking a cruise when the ship begins to sink. The life rafts hold heavy containers of survival supplies. Your friend says, "We need to tie ropes to the containers, toss them overboard, and let them hang submerged below the raft. That will allow the raft to carry more people without sinking." In a clear, coherent argument, explain whether your friend's idea will allow the raft to carry more people without sinking. Hurry and answer—the ship is sinking!

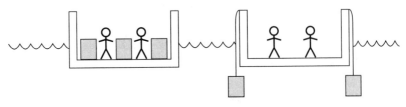

26. A raft of length x and width y floats partially submerged a distance z_2 in the ocean, as shown in the figure. The density of the ocean water is ρ. Atmospheric pressure is P_{atm}.

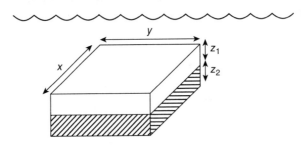

(A) Write an algebraic equation for the water pressure exerted on the bottom of the raft.
(B) Write an algebraic equation for the net buoyancy force the water exerts on the bottom of the raft.
(C) Write an algebraic equation for the amount of mass (M) that can be loaded onto the raft without it sinking.

Questions 27–29

A string holds a 10-cm × 10-cm × 20-cm block of wood with a density of 600 kg/m³ underwater, as shown in the figure.

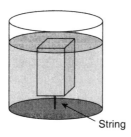

27. Draw a free body diagram, making sure that all the forces are drawn to relative scale.

28. Calculate the tension in the string.

29. Imagine the string is cut. Calculate the initial acceleration of the block of wood.

30. A hose with a 2-cm radius supplies water to fill a pool. Water flows out of the hose at a rate of 6 m/s. The pool has a length of 12 m, a width of 9 m, and a depth of 2 m.
    (A) What is the volume flow rate of the water exiting the hose?
    (B) How many hours does it take to completely fill the pool?

31. A river narrows as it passes through a canyon.
    (A) What happens to the volume flow rate of the water in the river as it passes through the canyon? Explain using a conservation law.
    (B) What happens to the speed of the water in the river as it passes through the canyon? Explain using a conservation law.

32. You visit a store and see a beach ball floating above a fan as shown in the figure. You notice that when the beach ball begins moving out of the air stream, it is pushed back toward the center of the stream. Explain why the balloon remains stable in the airflow from the fan.

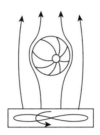

33. The figure shows air flowing around an aircraft wing. Using both the continuity equation and Bernoulli's equation, explain why this wing shape creates an upward force on the wing.

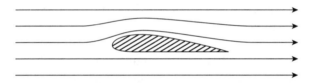

Questions 34 and 35

An open topped container is filled with water, as shown in the figure. A spigot is opened at the bottom of the container, and water flows out.

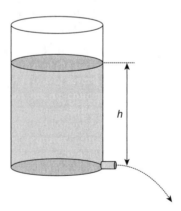

34. Use Bernoulli's equation to derive an equation for the exit velocity of the water.

35. Fill in the pressure-energy density bar chart for the fluid-earth system.

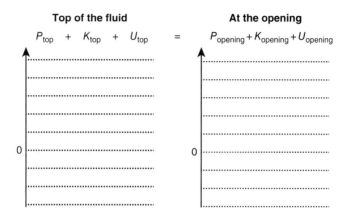

Questions 36 and 37

Fluid enters the left side of the flow tube and exits the right side at a higher velocity, as predicted by the continuity equation.

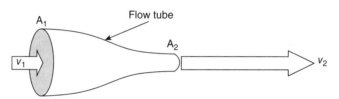

36. Fill in the pressure-energy density bar chart for the fluid-earth system, and use it to explain changes in the fluid pressure from side 1 to side 2 of the tube.

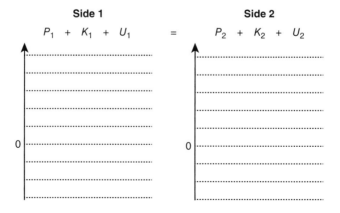

37. There are bubbles in the fluid as it flows from side 1 to side 2. Explain what happens to the size of the bubbles as they move toward side 2.

Questions 38 and 39

Water flows at 8 m/s from two 3-cm-diameter pipes into one 6-cm-diameter pipe.

38. Does the water speed up or slow down upon entering the 6-cm pipe? Justify your answer.

39. Calculate the speed of the water in the 6-cm pipe.

40. Water passes through a closed piping system starting at point 1 and exits to the atmosphere at the highest point near point 4. The pipe is small at point 1 but then widens out to a constant radius for the rest of the pipe.

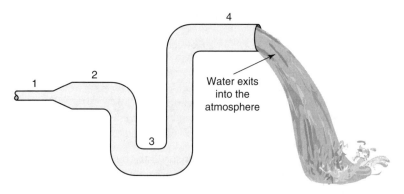

(A) Rank the velocity of the water in the pipe at the four locations from greatest to least. Justify your answer.
(B) Rank the pressure in the pipe at the four points from greatest to least. Justify your answer.

41. A variable-diameter pipe is used to supply water to a sprinkler system. Which of the sprinklers shown in the figure will squirt water the highest? Justify your claim.

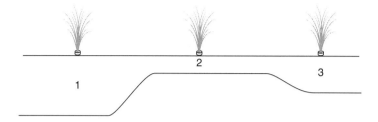

42. A dam controls the water level of the lake behind it by opening a flow pipe to allow water to exit out the base of the dam. The water level is 10 m above the pipe inlet. The pipe descends 20 m to the exit on the other side of the dam. The pipe diameters are shown in the figure below. Calculate the exit velocity of the water from the dam. Assume water density = 1,000 kg/m³, and atmospheric pressure is 100,000 Pa.

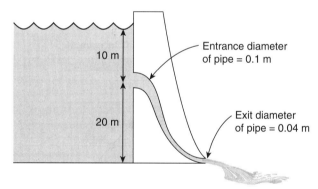

AP-Style Multiple-Choice Questions

43. Two blocks of different sizes and masses float in a tray of water. Each block is half submerged, as shown in the figure. Water has a density of 1,000 kg/m³. What can be concluded about the densities of the two blocks?

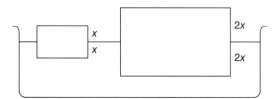

(A) The two blocks have different densities, both of which are less than 1,000 kg/m³.
(B) The two blocks have the same density of 500 kg/m³.
(C) The two blocks have the same density, but the density cannot be determined with the information given.
(D) The larger block has a greater density than the smaller block, but the densities of the blocks cannot be determined with the information given.

44. The figure shows four cylinders of various diameters filled to different heights with water. A hole in the side of each cylinder is plugged by a cork. All cylinders are open at the top. The corks are removed and the velocity of the water exiting the holes is measured before the water level has a chance to change. Which of the following is the correct ranking of the velocity of the water (v) as it exits each cylinder?

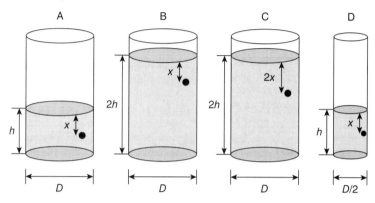

(A) $v_A > v_D > v_C > v_B$
(B) $v_A = v_D > v_C > v_B$
(C) $v_B > v_C > v_A = v_D$
(D) $v_C > v_A = v_B = v_D$

Questions 45 and 46

Four differently shaped sealed containers are completely filled with alcohol, as shown in the figure. Containers A and B are cylindrical. Containers C and D are truncated conical shapes. The top and bottom diameters of the containers are shown.

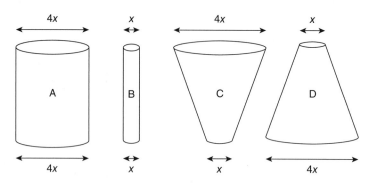

45. Which of the following is the correct ranking of the pressure (P) at the bottom of the containers?

    (A) $P_A = P_B = P_C = P_D$
    (B) $P_A = P_D > P_C = P_B$
    (C) $P_A > P_D > P_C > P_B$
    (D) $P_D > P_A > P_C > P_B$

46. The force on the bottom of container A due to the fluid inside the container is F. What is the force on the bottom of container B due to the fluid inside?

    (A) F
    (B) $F/4$
    (C) $F/8$
    (D) $F/16$

47. Two cylinders filled with a fluid are connected by a pipe so that fluid can pass between the cylinders, as shown in the figure. The cylinder on the right has 4 times the diameter of the cylinder on the left. Both cylinders are fitted with a movable piston and a platform on top. A person stands on the left platform. Which of the following lists the correct number of people that need to stand on the right platform so neither platform moves? Assume that the platform and piston have negligible mass and that all the people have the same mass.

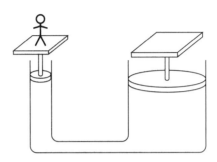

(A) 16 people
(B) 4 people
(C) 1 person
(D) It is impossible to balance the system because you need 1/16 of a person on the right side.

48. A mass (m) is suspended in a fluid of density (ρ) by a thin string, as shown in the figure. The tension in the string is T. Which of the following is an appropriate equation for the buoyancy force? *Select two answers.*

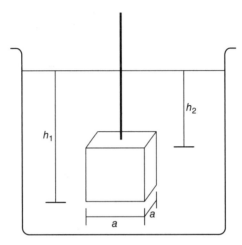

(A) $F_b = mg$
(B) $F_b = mg - T$
(C) $F_b = a^2 \rho g h_1$
(D) $F_b = a^2 \rho g (h_1 - h_2)$

49. Three wooden blocks of different masses and sizes float in a container of water, as shown in the figure. Each of the masses has a weight on top. Which of the following correctly ranks the buoyancy force on the wooden blocks?

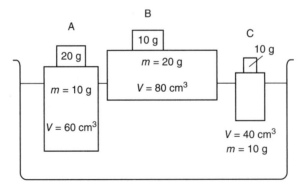

(A) $A > B = C$
(B) $A = B > C$
(C) $B > A = C$
(D) $B > A > C$

50. Two blocks of the same dimensions are floating in a container of water, as shown in the figure. Which of the following is a correct statement about the two blocks?

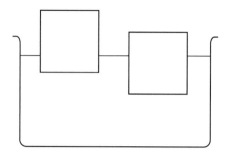

(A) The net force on both blocks is the same.
(B) The buoyancy force exerted on both blocks is the same.
(C) The density of both blocks is the same.
(D) The pressure exerted on the bottom of each block is the same.

51. The figure shows four cubes of the same volume at rest in a container of water. Cube C is partially submerged. Cubes A, B, and D are fully submerged, with B resting on the bottom of the container. Which of the following correctly ranks the densities (ρ) of the cubes? Assume the water to be incompressible.

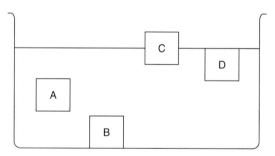

(A) $\rho_C > \rho_D > \rho_A > \rho_B$
(B) $\rho_B > \rho_A > \rho_D > \rho_C$
(C) $\rho_B > \rho_A = \rho_D > \rho_C$
(D) $\rho_B > \rho_A = \rho_D = \rho_C$

52. A beaker of water sits on a balance. A metal block with a mass of 190 g is held suspended in the water by a spring scale in position 1, as shown in the figure. In this position, the reading on the balance is 1,260 g, and the spring scale reads 120 g. When the block is lifted from the water to position 2, what are the readings on the balance and spring scale?

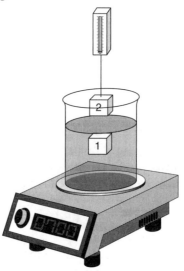

	Balance reading	Spring scale reading
(A)	1,190 g	120 g
(B)	1,190 g	190 g
(C)	1,260 g	190 g
(D)	1,260 g	120 g

53. Blood cells pass through an artery that has a buildup of plaque along both walls, as shown in the figure. Which of the following correctly describes the behavior of the blood cells as they move from the right side of the figure through the area of plaque? Assume the blood cells can change volume.

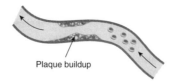

Plaque buildup

(A) The blood cells increase in speed and expand in volume.
(B) The blood cells increase in speed and decrease in volume.
(C) The blood cells decrease in speed and expand in volume.
(D) The blood cells decrease in speed and decrease in volume.

54. Firefighters use a hose with a 2-cm diameter exit nozzle connected to a hydrant with an 8-cm-diameter opening to attack a fire on the second floor of a building 6 m above the hydrant, as shown in the figure. What pressure must be supplied at the hydrant to produce an exit velocity of 18 m/s? (Assume the density of water is 1,000 kg/m³, and the exit pressure is 1×10^5 Pa.)

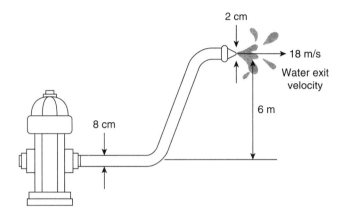

(A) 1.7×10^5 Pa
(B) 2.0×10^5 Pa
(C) 2.6×10^5 Pa
(D) 3.2×10^5 Pa

55. A 1-cm-diameter pipe leads to a showerhead with twenty 1-mm diameter exit holes. The velocity of the water in the pipe is v. What is the velocity of the water exiting the holes?
(A) $0.05\ v$
(B) $0.5\ v$
(C) $5\ v$
(D) $100\ v$

AP-Style Free-Response Questions

56. An air bubble is released from the bottom of a swimming pool and ascends to the surface.
    (a) In a clear, coherent, paragraph-length response, describe any changes in the bubble size and describe the motion of the bubble as it ascends to the surface. Explain the factors that affect the size of the bubble and the bubble's motion. Include a description of any forces acting on the bubble from the time it is at the bottom of the pool until it reaches the surface.
    (b) Draw a diagram of all the forces acting on the bubble. Make sure the forces are in correct proportion.
    (c) The bubble does not collapse under the pressure of the water. Explain how the behavior of the gas atoms keep the bubble from collapsing.
    (d) Assume that the air temperature inside the bubble remains constant as it rises. The bubble has an initial volume of V_D, begins at a depth of D below the surface of the water, and reaches the surface where the pressure is P_S. The density of the water is ρ.
        i. Derive an expression for the initial pressure (P_D) in the bubble in terms of the given quantities and known constants.
        ii. Derive an expression for the volume (V_S) of the bubble when it reaches the surface.
    (e) The temperature of the bubble is measured as it travels upward. It is found that the assumption of constant temperature in (d) is incorrect. The temperature of the air actually decreases as it rises. Now assume that the bubble rises so quickly that there is negligible thermal energy transfer between the bubble and the swimming pool. Base your answers on this new assumption.

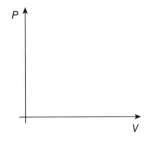

        i. Sketch the process on the PV diagram. Indicate on the axis the initial and final pressures and volumes.

ii. How does the value $P_S V_S$ compare to the value $P_D V_D$?
___Greater than $P_D V_D$ ___Equal to $P_D V_D$ ___Less than $P_D V_D$
Justify your answer.

57. A 1.0-cm-radius hose with a 0.50-cm-radius exit nozzle is being used to fill a 1,000-ml beaker with oil (1,000 ml = 0.0010 m³). The velocity of the oil in the hose is $v = 0.40$ m/s as shown in the figure. The density of the oil is 960 kg/m³, and the atmospheric pressure is 1.01×10^5 Pa.

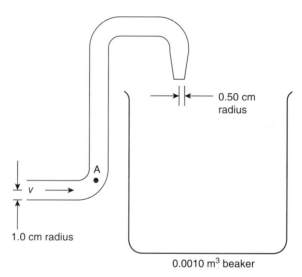

(a) The nozzle attached to the end of the hose has a smaller radius than the hose. If the nozzle is removed from the hose, will the beaker be filled faster? Justify your answer with a physics principle or law.
(b) Calculate the exit velocity of the oil from the nozzle.
(c) How long will it take to fill the beaker?
(d) Point A is shown in the figure. How does the pressure in the fluid at point A compare to the pressure in the fluid at the exit nozzle? Justify your claim.
(e) The hose is now used to fill a 200-ml graduated cylinder with oil to the same height as the height of the oil in the 1,000-ml beaker. Compare the net force from the oil on the bottom of the 200-ml cylinder and the 1,000-ml beaker. Explain your answer.

(f) A cube of lead with a side dimension of 5.0 cm is slowly lowered into the beaker of oil by a thin string attached to a spring scale at a constant rate, as shown in the figure. The density of lead is 11,300 kg/m³.

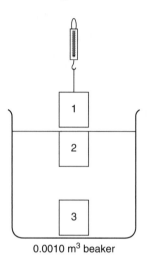

0.0010 m³ beaker

  i. What will be the spring scale reading in newtons when the lead has been submerged to location 2?
 ii. Does the spring scale reading increase, decrease, or stay the same when the cube is lowered from location 2 to location 3? Justify your answer by referencing the pressure of the fluid on the lead cube.
iii. The lead cube is lowered from above the oil's surface (location 1) to a spot just below the surface (location 2) until the cube is just above the bottom of the beaker (location 3). Describe any changes in pressure on the bottom of the beaker during this process. Explain your answer.

CHAPTER 2

Thermodynamics and Gases

Skill-Building Questions

58. Explain the difference and similarities between gases and liquids.

59. Use the microscopic motion of atoms to describe how gases exert pressure on a surface.

60. Gas pressure always creates a force that is perpendicular to the surface with which the gas is in contact. Explain why this is true.

61. Briefly explain how a suction cup works.

62. What are the basic assumptions of the Ideal Gas model?

63. Explain what temperature really measures.

64. Describe any changes in the kinetic energy of a gas as the gas temperature changes from T to T/2.

65. Describe the changes in the motion of the gas molecules as the gas temperature changes from T to T/2.

66. Use the axis to sketch the speed distributions listed below.

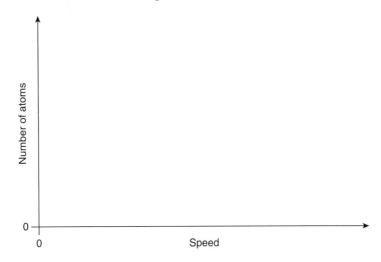

(A) Sketch the speed distribution of 2 moles of neon gas at a temperature of 300 K. The horizontal axis represents the speed of the ^{20}Ne atoms. The vertical axis designates the number of neon atoms moving at a particular speed.
(B) Sketch the speed distribution of 2 moles of helium at a temperature of 300 K. Sketch the ^{4}He distribution in correct relationship to the ^{20}Ne distribution.
(C) Sketch the speed distribution of 2 moles of argon at a temperature of 300 K. Sketch the ^{40}Ar distribution in correct relationship to the ^{20}Ne and ^{4}He distributions.

67. Carbon monoxide gas with a molecular mass of 28.0 kg/kmol and an initial temperature of 200 K is confined to the left side of a sealed container. Diatomic nitrogen gas with a molecular mass of 28.0 kg/kmol and an initial temperature of 400 K is confined to the right side of the sealed container, as shown in the figure. Separating the gases is a removable barrier. When the barrier is removed, the two gases mix.

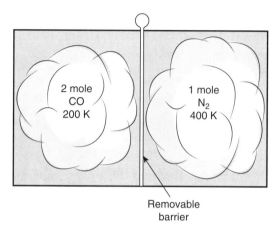

(A) Describe the temperature changes in the two gases.
(B) Describe any movement of thermal energy over a long period of time. Explain the microscopic process that determines this process.
(C) Sketch the initial molecular speed distributions of the two gases on the axis. Label each gas. Explain any differences in the speed distribution of the two gases.

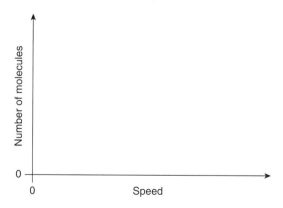

(D) Sketch the final molecular speed distributions of the two gases on the axis. Label each gas. Explain the changes in the graph from the initial condition. Also explain any differences in the speed distributions of the two gases.

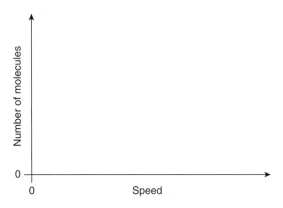

68. Your physics teacher instructs you to determine the relationship between gas pressure and volume.
    (A) List the items you would use to perform this investigation.
    (B) Sketch a simple diagram of your investigation. Make sure to label all items, and label any measurements that will be made.
    (C) Outline the experimental procedure you would use to gather the necessary data. Indicate the measurements to be taken, and how the measurement will be used to obtain the data needed. Make sure your outline contains sufficient detail so that another student could follow your procedure and duplicate your results.
    (D) On the axis, sketch the line or curve that you predict will represent a plot of the data gathered from this experiment.

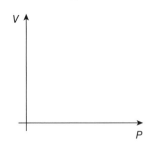

69. A physicist is designing an experiment to determine the relationship between the volume of a gas and the temperature of a gas.

(A) List the items the physicist could use to perform this investigation.

(B) Sketch a simple diagram of the investigation. Make sure to label all items used in the sketch and label any measurements that will be made.

(C) Outline the experimental procedure the physicist could use to gather the necessary data. Indicate the measurements to be taken, and how the measurement will be used to obtain the data needed. Make sure your outline contains sufficient detail so that another scientist could follow your procedure and verify your results.

(D) Explain how data from this experiment could be used to determine the value of absolute zero.

(E) On the axis, sketch the line or curve that you predict will represent the relationship between volume and temperature as shown in the data gathered in this experiment.

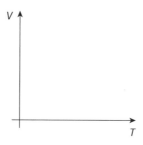

Questions 70–73

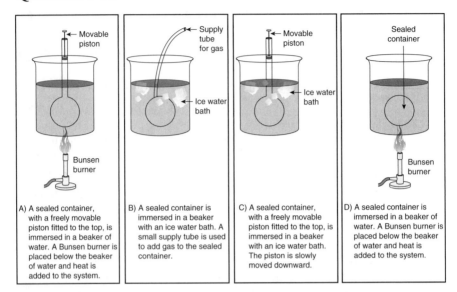

70. Which of the experimental setups shown in the figure above is most likely to produce the graph shown?

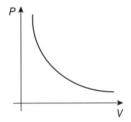

(A) A
(B) B
(C) C
(D) None of the experimental setups will produce this graph.

71. Which of the experimental setups shown in the figure on page 58 is most likely to produce the graph shown?

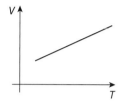

(A) A
(B) None of the experimental setups will produce this graph.
(C) C
(D) D

72. Which of the following data was most likely produced by experiment D shown on page 58?

(A)

P (kPa)	V (m³)
100	0.25
150	0.17
180	0.14
220	0.11
260	0.096

(B)

T (K)	V (m³)
280	0.0120
290	0.0124
340	0.0146
350	0.0150
360	0.0154

(C)

T (K)	P (kPa)
280	100
300	107
320	114
340	121
360	129

(D)

T (K)	n (mol)
100	2.00
150	1.33
200	1.00
250	0.80
300	0.67

73. Which of the following graphs best represents the results of experiment B shown in the figure on page 58?

(A)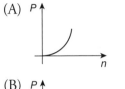

(C)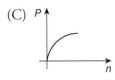

(B)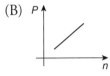

(D)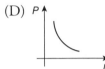

74. In an experiment, a gas is confined in a cylinder with a movable piston. Force is applied to the piston to increase the pressure and change the volume of the gas. Each time the gas is compressed, it is allowed to return to a room temperature of 20°C. The experimental data is shown in the table. Calculate the number of molecules and moles of gas in the cylinder.

Pressure ($\times 10^5$ Pa)	Volume ($\times 10^{-3}$ m^3)
1.0	25
1.5	17
1.8	14
2.2	11
2.6	9.6
3.3	7.6

75. Explain how the thermal energy of an ideal gas can be calculated from the temperature. What, if any, additional information will be needed?

76. Why does the internal energy of a gas not include any potential energy?

77. Use the figure to answer the following questions.

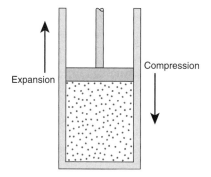

(A) Explain why work done on the gas is negative when the gas is expanding and positive when the gas is being compressed.
(B) Explain why doing work on the gas by pushing the piston to compress the gas increases the molecular kinetic energy of the gas.
(C) Use the behavior of atoms to explain why a gas will lose internal energy when the gas expands by moving a piston upward.

78. Design an experiment to determine the work done on a gas by an external force.
(A) List the items you would use to perform this investigation.
(B) Sketch a simple diagram of your investigation. Make sure to label all items and label any measurements that will be made.
(C) Outline the experimental procedure you would use to gather the necessary data. Indicate the measurements to be taken and how the measurement will be used to obtain the data needed. Make sure your outline contains sufficient detail so that another student could follow your procedure and duplicate your results.
(D) On the axis, sketch the line or curve that you predict will represent a plot of the data gathered from this experiment. Show how the graph could be used to determine the work done on the gas.

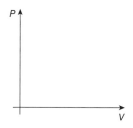

79. What are the two ways to change the energy of a system? Explain them both.

80. What are the three processes by which heat can be transferred from one system to another? Describe each in detail, being sure to reference the atomic nature of matter.

Questions 81–83

You have the job of finding out how multiple factors influence the rate at which thermal energy is conducted through a cylindrical rod. You are given the following equipment:
- Copper rods 1.0 cm in diameter and 10 cm, 20 cm, 30 cm, and 50 cm in length
- Aluminum rod 1.0 cm in diameter and 20 cm in length
- Steel rods 20 cm in length with a square cross-sectional area; 0.25 cm, 0.50 cm, 0.75 cm, and 1.0 cm thickness
- Temperature sensor
- Wax
- Bunsen burner
- Hot plate
- Ice
- Water
- Beaker
- Stopwatch
- Other standard lab equipment

81. The first investigation is to determine the relationship between the path length and the rate of thermal energy transfer.
    (A) List the items you would use to perform this investigation.
    (B) Outline the experimental procedure you would use to gather the necessary data. Indicate any measurements taken, and how the measurement will be used. Make sure your outline contains sufficient detail so that another student could follow your procedure and duplicate your results.

(C) On the axis, sketch the relationship you expect to find from this investigation. Explain why you expect this relationship.

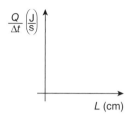

82. The second investigation is to determine the relationship between the temperature difference and the rate of thermal energy transfer.
    (A) List the items you would use to perform this investigation.
    (B) Outline the experimental procedure you would use to gather the necessary data. Indicate any measurements taken, and how the measurement will be used. Make sure your outline contains sufficient detail so that another student could follow your procedure and duplicate your results.
    (C) On the axis, sketch the relationship you expect to find from this investigation. Explain why you expect this relationship.

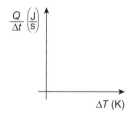

83. The third and final investigation is to determine the relationship between the cross-sectional area of contact and the rate of thermal energy transfer.
    (A) List the items you would use to perform this investigation.
    (B) Outline the experimental procedure you would use to gather the necessary data. Indicate any measurements taken, and how the measurement will be used. Make sure your outline contains sufficient detail so that another student could follow your procedure and duplicate your results.

(C) On the axis, sketch the relationship you expect to find from this investigation. Explain why you expect this relationship.

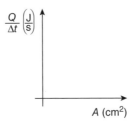

84. A small spoon is placed in a large, hot cup of coffee. Heat flows from the coffee to the spoon until both reach the same temperature.
    (A) Explain why heat flows into the spoon from the coffee and not the other way around.
    (B) Explain why heat transfer ceases when the spoon and the coffee reach the same temperature.
    (C) At thermal equilibrium, do the spoon and the coffee have equivalent thermal energies? Justify your answer.
    (D) Describe the changes to the entropy of the spoon and coffee, as well as the net entropy of the spoon-coffee system during this process. Explain your reasoning.

85. What do *reversible* and *irreversible* mean?

Questions 86–89

For each of the situations described below:
- Sketch the path of the process for each gas on the *PV* diagram beginning at point 1.
- Complete the energy bar chart of the gas for each process.

86.

Description	PV diagram	Energy bar chart
A sealed rigid container filled with 600-K air is completely immersed in an ice bath. The container remains in the ice bath until the gas reaches a final temperature of 200 K.	P ($\times 10^5$ Pa) vs V ($\times 10^{-4}$ m^3); point at (6, 3) labeled 1	

87.

Description	PV diagram	Energy bar chart
A cylinder, fitted at the top with a movable piston with a mass of 200 g, is filled with 300-K nitrogen. The bottom of the container is held over a Bunsen burner. The piston rises. This is continued until the final temperature of the gas reaches 600 K.	P ($\times 10^5$ Pa) vs V ($\times 10^{-4}$ m^3); point at (6, 3) labeled 1	

88.

Description	PV diagram	Energy bar chart
A cylinder filled with air is fitted at the top with a movable piston. Small 20-g masses are added, one at a time, on top of the piston. After each mass is added, the air inside the cylinder is allowed to reach thermal equilibrium with the environment before the next mass is added. This continues until the gas is one-third the original volume.	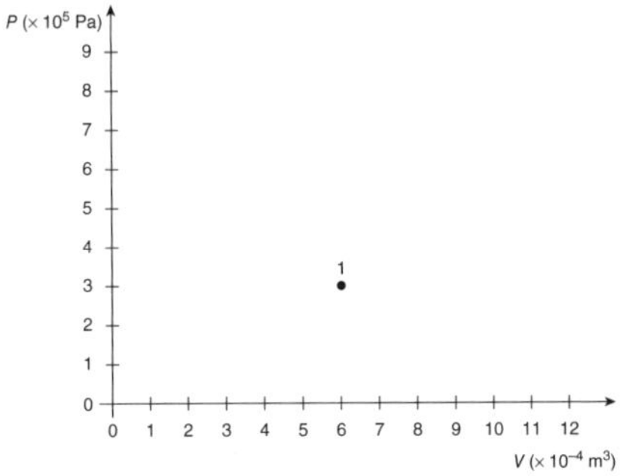 P ($\times 10^5$ Pa) vs V ($\times 10^{-4}$ m³); point plotted at (7, 3).	

89.

Description	PV diagram	Energy bar chart
A bicycle pump has a cylinder filled with air fitted at the top with a movable piston and handle. A student seals the exit valve so the air cannot escape; takes the handle; and quickly pushes down, decreasing the air volume by half. The process is so quick that no thermal energy escapes from the gas to the environment.	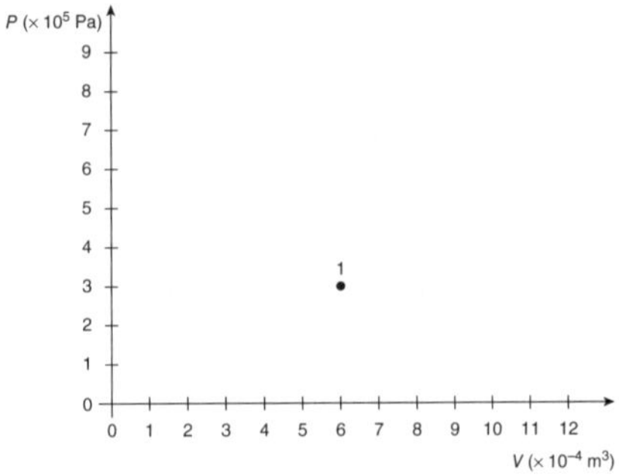 P ($\times 10^5$ Pa) vs V ($\times 10^{-4}$ m³); point plotted at (7, 3).	

90. Three identical blocks of differing temperatures are stacked on top of one another and insulated from the environment, as shown in the figure.

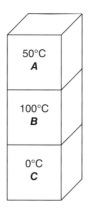

(A) Plot the temperature of the blocks versus time. Label each line and indicate all important temperatures on the graph.

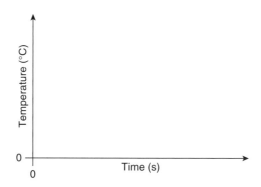

(B) Discuss any changes in entropy for block *C*. Explain your answer at the microscopic level.//
(C) Discuss the entropy of the system consisting of all three blocks from initial to final state. Justify your answer.

91. A gas moves through a process shown in the figure. Is the process shown isothermal? Explain.

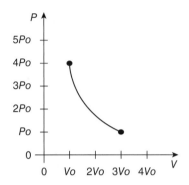

92. The figure shows a sample of gas that is taken through four stages. Each stage is labeled. Explain which of the labeled points has the highest and lowest temperature.

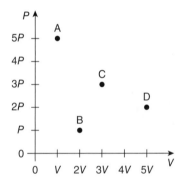

93. The figure shows the pressure and volume of a gas sample. Can you determine the temperature of the sample? If you can, calculate it. If not, what additional information is needed?

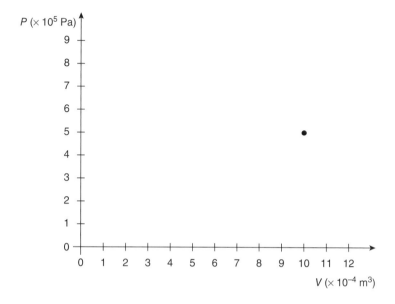

94. A thermodynamic process is shown in the figure. Sketch a new process that starts at the same initial state of pressure and volume and that displays the following characteristics.

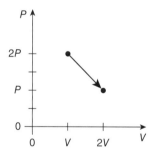

(A) Expands from V to $2V$ and has a lower final temperature than the process shown.
(B) Expands from V to $2V$ along a different path while still having the same magnitude of work as the process shown.
(C) Expands from V to $2V$ along a path so that there is more heat being added to the gas than the process shown.

95. Explain what is happening to the entropy of the gas as it moves through the process from point A to point B and from point B to point C, as shown in the figure.

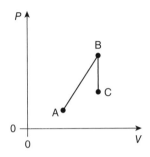

96. A gas is taken through the four processes shown in the figure. For each path, identify the name of the process and determine if the values of ΔT, ΔU, W, and Q are positive, negative, or zero and fill in the table with a +, −, or 0. Explain your reasoning.

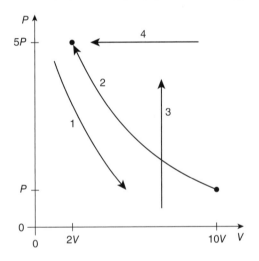

The table is started for you! Process #1 is adiabatic.

Name of process	ΔT	ΔU	W	Q
Adiabatic				

97. Two moles of a gas are taken through the thermodynamic process ABCA, as shown in the figure.

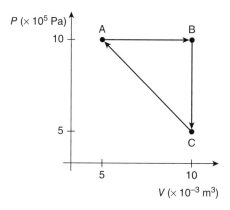

(A) Rank the work in the steps of the process from most positive to most negative.
(B) Rank the change in temperature of the gas for each step in the process from most positive to most negative.
(C) Rank the thermal energies of the points A, B, and C from greatest to least.
(D) Calculate the temperature of point B.
(E) Calculate the change in temperature for the entire cycle ABCA.
(F) Calculate the work done in process C→A.
(G) Calculate the heat flow for process C→A.

Questions 98–100

For each of the following data sets, explain which thermodynamic process is represented. Defend your answer.

98.

Pressure (kPa)	Volume (ml)
120	60
140	60
155	60
180	60
230	60

99.

Pressure (kPa)	Volume (ml)
101	12
101	24
101	36
101	44
101	60

100.

Pressure (kPa)	Volume (ml)
50	120
100	60
200	30
400	15

AP-Style Multiple-Choice Questions

101. Air is made up primarily of nitrogen and oxygen. In an enclosed room with a constant temperature, which of the following statements is correct concerning the nitrogen and oxygen gases in the air?

(A) The nitrogen gas molecules have a higher average kinetic energy than the oxygen gas molecules.

(B) The nitrogen gas molecules have the same average kinetic energy as the oxygen gas molecules.

(C) The nitrogen gas molecules have a lower average kinetic energy than the oxygen gas molecules.

(D) More information is necessary to compare the average kinetic energies of the two gases.

102. Air is made up primarily of nitrogen and oxygen. In an enclosed room with a constant temperature, which of the following statements is correct concerning the nitrogen and oxygen gases in the air?
    (A) The nitrogen gas molecules have a higher velocity than the oxygen gas molecules.
    (B) The nitrogen gas molecules have the same velocity as the oxygen gas molecules.
    (C) The nitrogen gas molecules have a lower velocity than the oxygen gas molecules.
    (D) It is impossible to compare the velocity of the two gases without knowing the temperature of the air and the percentage of nitrogen and oxygen in the room.

103. In an experiment, a gas is confined in a cylinder with a movable piston. Force is applied to the piston to increase the pressure and change the volume of the gas. Each time the gas is compressed, it is allowed to return to a room temperature of 20°C. The data gathered from the experiment is shown in the table. What should be plotted on the vertical and horizontal axes so the slope of the graph can be used to determine the number of moles of gas in the cylinder?

Pressure ($\times 10^5$ Pa)	Volume ($\times 10^{-3}$ m^3)
1.0	25
1.5	17
1.8	14
2.2	11
2.6	9.6
3.3	7.6

(A) P and V^2
(B) P and V
(C) P and $(V)^{1/2}$
(D) P and $1/V$

104. In an experiment, a sealed container with a volume of 100 ml is filled with hydrogen gas. The container is heated to a variety of temperatures, and the pressure is measured. The data from the experiment is plotted in the figure. Which of the following methods can be used to determine additional information regarding the gas? *Select two answers.*

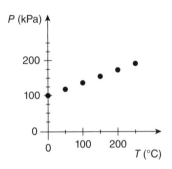

(A) The slope can be used to calculate the number of atoms in the gas.
(B) The area under the graph can be used to calculate the work done on the gas.
(C) The vertical axis can be used to calculate the force the gas exerts on the container.
(D) The x-intercept can be used to estimate the value of absolute zero.

105. Two identical rooms are connected by an open door. The temperature in one room is greater than the temperature in the other. Which room contains the most gas molecules?
(A) The warmer room.
(B) The colder room.
(C) The number of gas molecules will be the same in both rooms.
(D) It is impossible to determine without more information.

106. On a hiking trip in the mountains, where the air temperature is cool and has a lower concentration of oxygen, you seal an empty water bottle. You return to your home near sea level where the air temperature is warm and has a higher concentration of oxygen. You notice that the sealed bottle appears partially crushed. Which of the following would contribute to the decrease in volume of the bottle?

(A) The change in temperature
(B) The change in atmospheric pressure
(C) The change in oxygen concentration
(D) The change in temperature, pressure, and oxygen concentration

107. The figure shows the pressure and volume of a gas at four different states. Which of the following correctly ranks the temperature of the gas at the different states?

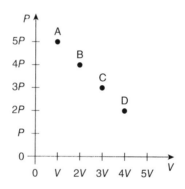

(A) $T_A > T_B > T_C > T_D$
(B) $T_B = T_C > T_A = T_D$
(C) $T_C > T_B = T_D > T_A$
(D) $T_D > T_C > T_B > T_A$

108. Which of the following is correct concerning the two processes shown in the figure?

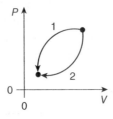

(A) $\Delta U_1 = \Delta U_2$ and $W_1 = W_2$
(B) $\Delta U_1 = \Delta U_2$ and $W_1 > W_2$
(C) $\Delta U_1 > \Delta U_2$ and $W_1 = W_2$
(D) $\Delta U_1 > \Delta U_2$ and $W_1 > W_2$

109. The figure shows four samples of gas being taken through four different processes. Process 1 is adiabatic. In which process is heat being transferred to the gas sample from the environment?

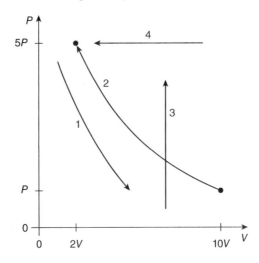

(A) 1
(B) 2
(C) 3
(D) 4

110. Two sealed cylinders holding different gases are placed one on top of the other so heat can flow between them. Cylinder A is filled with hydrogen. Cylinder B is filled with helium moving with an average speed that is half that of the hydrogen atoms. Helium atoms have four times the mass of hydrogen atoms. Which of the following best describes the transfer of heat between the two containers by conduction?

   (A) Net heat flows from cylinder A to cylinder B, because heat flows from higher kinetic energy atoms to lower kinetic energy atoms.
   (B) Net heat flows from cylinder B to cylinder A, because heat flows from higher kinetic energy atoms to lower kinetic energy atoms.
   (C) There is no net heat transfer between the two cylinders, because both gases have the same average atomic kinetic energy.
   (D) There is no net heat transfer between the two cylinders, because heat conduction requires the movement of atoms between the cylinders, but the cylinders are sealed.

Questions 111 and 112

A gas beginning at point O on the graph can be taken along four paths to different ending conditions.

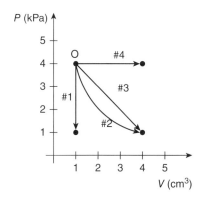

111. Which of the following are the same for processes 2 and 3? *Select two answers.*

   (A) Q
   (B) ΔT
   (C) ΔU
   (D) W

112. Along which of the paths is the most thermal energy removed from the gas?
    (A) 1
    (B) 2
    (C) 3
    (D) 4

113. The graph shows the distribution of speeds for one mole of hydrogen at temperature T, pressure P, and volume V. How would the graph change if the sample was changed from one mole hydrogen to one mole of argon at the same temperature, pressure, and volume?

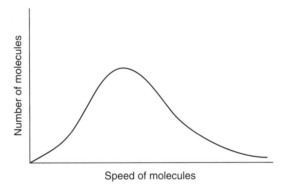

   (A) The peak will shift to the left
   (B) The peak will shift upward and to the left
   (C) The peak will shift to the right
   (D) The peak will shift downward and to the right

114. The graph shows the pressure and volume of a gas being taken from state #1 to state #2. Which of the following correctly indicates the sign of the work done by the gas, and the change in temperature of the gas?

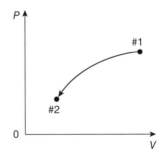

	Work done	Δ Temperature
(A)	+	+
(B)	+	−
(C)	−	+
(D)	−	−

Questions 115 and 116

A resistor of resistance (*R*) is sealed in a closed container with *n* moles of gas inside. A battery of emf (ε) is connected to the resistor.

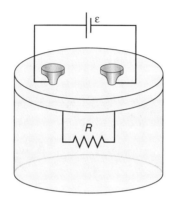

115. Which of the following graphs shows the correct relationship between the gas atoms' average velocity (v_{avg}) and electrical energy (*E*) supplied to the resistor?

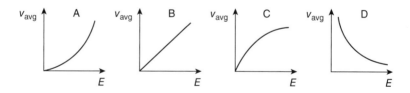

116. Which of the following graphs shows the correct relationship between gas pressure (*P*) and electrical energy (*E*) supplied to the resistor?

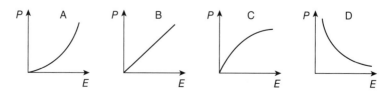

AP-Style Free-Response Questions

117. A mole of ideal gas is enclosed in a cylinder with a movable piston with a cross-sectional area of 1×10^{-2} m². The gas is taken through a thermodynamic process, as shown in the figure.

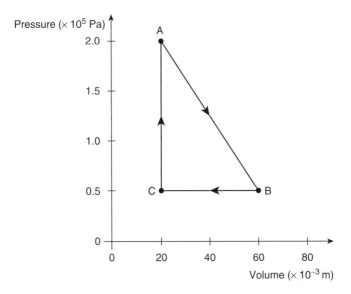

(a) Calculate the temperature of the gas at state A, and describe the microscopic property of the gas that is related to the temperature.
(b) Calculate the force of the gas on the piston at state A, and explain how the atoms of the gas exert this force on the piston.
(c) Predict qualitatively the change in the internal energy of the gas as it is taken from state B to state C. Justify your prediction.
(d) Is heat transferred to or from the gas as it is taken from state B to state C? Justify your answer.
(e) Discuss any entropy changes in the gas as it is taken from state B to state C. Justify your answer.
(f) Calculate the change in the total kinetic energy of the gas atoms as the gas is taken from state C to state A.

(g) On the axis provided, sketch and label the distribution of the speeds of the atoms in the gas for states A and B.

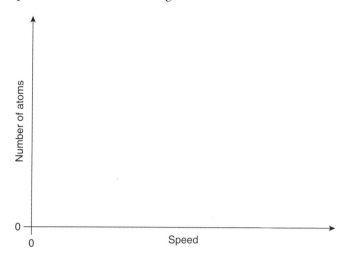

118. You wish to determine the relationship between gas pressure and temperature.
   (a) List the items you would use to perform this investigation.
   (b) Draw a simple picture of the lab setup, and outline the experimental procedure you would use to gather the necessary data. Indicate the measurements to be taken and how the measurement will be used to obtain the data needed. Make sure your outline contains sufficient detail so that another student could follow your procedure.
   (c) On the axis, sketch the line or curve that you predict will represent the results of the data gathered in this experiment.

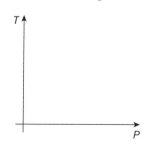

(d) Explain how you could use your results to estimate the value of absolute zero.

You are given the following set of data acquired in a gas laboratory experiment and asked to determine the relationship between pressures and volume for the gas.

Trial	Temperature (K)	Volume (ml)	Pressure (kPa)
1	270	500	8,979
2	270	1,000	4,490
3	270	2,000	2,245
4	270	5,000	898
5	300	1,000	4,988
6	300	2,000	2,494
7	300	5,000	998
8	320	1,000	5,321
9	320	5,000	1,064
10	350	500	11,640
11	350	5,000	1,164
12	370	500	12,305
13	370	2,000	3,076
14	370	3,000	1,230
15	370	6,000	769

(e) Which subset of the data would be most useful in creating a graph to determine the relationship between gas pressure and volume? Explain why the trials you selected are the most useful.

(f) Plot the subset of data you chose on the graph, being sure to label the axes. Draw a line or curve that best represents the relationship between the variables.

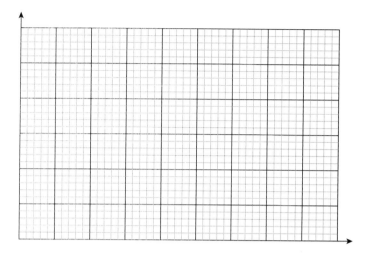

(g) What can you conclude from your line or curve about the relationship between volume and pressure?

CHAPTER 3

Electric Force, Field, and Potential

Skill-Building Questions

Questions 119 and 120

A triboelectric series is given for reference.

Triboelectric series
Increasingly positive
↑ Human skin
Rabbit fur
Glass
Human hair
Silk
Paper
Cotton
Zero affinity
Steel
Wood
Polystyrene (Styrofoam)
Polyethylene (clear sticky tape)
Rubber balloon
Polyester
PVC
↓ Kitchen plastic wrap (Saran)
Increasingly negative

119. You notice that when a polyester shirt rubs against your skin it starts to cling to your skin. Use the molecular scale to explain this phenomenon.

120. You rub a balloon on your hair and then suspend the balloon from the ceiling by a thread. For each of the following objects that are brought near the suspended balloon, describe what will happen to the balloon and explain the behavior.
    (A) Glass rubbed with kitchen plastic wrap
    (B) A PVC pipe rubbed with silk
    (C) A piece of uncharged paper

121. The bottom of a metal pie pan is filled with puffed rice cereal.
    (A) A PVC pipe that has been rubbed with a hand is touched to the pie pan. Explain what, if anything, happens with the cereal.
    (B) A glass rod rubbed with kitchen plastic wrap is held above and near but not touching the cereal. Some of the cereal flies from the pan toward the glass rod. Why does this occur?
    (C) Some of the cereal that flies up toward the rod flies away after touching it. Explain this behavior in detail.

122. The north end of a compass is brought near a negatively charged balloon. Will the compass be affected by the balloon? Justify your claim.

123. (A) A negatively charged piece of clear sticky tape is brought near an aluminum can without touching. What, if anything, happens to the tape? Explain.
    (B) A negatively charged balloon that had a charge twice that of the clear sticky tape is brought near the aluminum can without touching. What, if anything, happens to the tape? Explain.

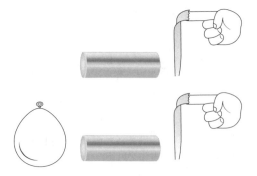

124. An aluminum leaf electroscope is used in physics labs to test the behavior of electric charges. An insulating rod is used to manipulate the leaves of the electroscope. In each of the following cases, do the leaves of the electroscope deflect more, deflect less, or remain the same? Explain your reasoning in each case.

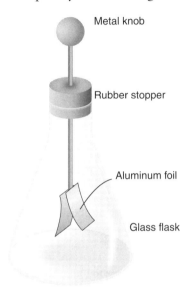

Initial charge of electroscope	Situation	Does the deflection of the leaves increase, decrease, or remain the same?	Justification for your answer
Neutral	Neutral rod touches the metal knob.		
Neutral	Positively charged rod touches the metal knob.		
Neutral	Positively charged rod is brought near but does not touch the metal knob.		

Initial charge of electroscope	Situation	Does the deflection of the leaves increase, decrease, or remain the same?	Justification for your answer
Positive	Neutral rod touches the metal knob.		
Positive	Neutral rod is brought near but does not touch the metal knob.		
Positive	Negatively charged rod touches the metal knob.		
Positive	Negatively charged rod is brought near but does not touch the metal knob.		
Negative	Negatively charged rod touches the metal knob.		
Negative	Negatively charged rod is brought near but does not touch the metal knob.		

125. Explain each of the following processes for charging an object.
    (A) Friction
    (B) Conduction
    (C) Induction

126. Your teacher gives you a charged metal sphere that rests on an insulating stand. The teacher asks you to determine if the charge on the object is positive or negative.
(A) List the items you would use to perform this investigation.
(B) Outline the experimental procedure you would use to make this determination. Indicate the measurements to be taken and how the measurements will be used to obtain the data needed. Make sure your outline contains sufficient detail so that another student could follow your procedure and duplicate your results.

127. A student claims that a charged Van de Graaff generator rearranges the electric charge in nearby conductors. How could you prove whether or not this claim is true? Design an investigation to test this claim.
(A) List the items you would use to perform this investigation.
(B) Outline the experimental procedure you would use to make this determination. Indicate the measurements to be taken and how the measurements will be used to obtain the data needed. Make sure your outline contains sufficient detail so that another student could follow your procedure and duplicate your results.

128. Explain what *grounding* means.

129. Explain what happens to the charge that is added to each of the following.
(A) A sphere made of copper
(B) A sphere made of rubber

130. The figure shows two conductive spheres (A and B) connected by a rod. Both spheres begin with no excess charge. A negatively charged rod is brought close to and held near sphere A as shown.

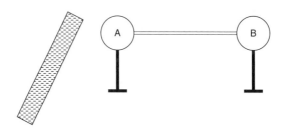

(A) If the connecting rod is made of wood, what is the net charge of the spheres while the rod is held in the position shown? Justify your answer.
(B) If the connecting rod is made of copper, what is the net charge of the spheres while the rod is held in the position shown? Justify your answer.
(C) The rod is now brought into contact with sphere A. How will this change the answers to the previous two questions? Explain.

131. A 5-g balloon is rubbed with rabbit fur and suspended from the ceiling by a light thread. An identically charged balloon is brought close so the centers of the balloons are 20 cm apart. This causes the suspended balloon to deflect outward so the thread makes an angle of 14° with the vertical, as shown in the figure.

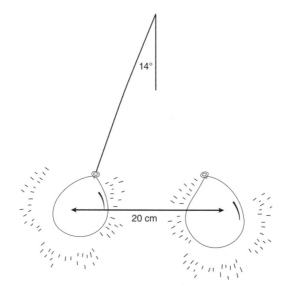

(A) Calculate the charge on the balloon. Show all your algebraic work in symbolic form. Include a force diagram of the balloon in your solution. List any assumptions you make to solve this problem.
(B) Determine the number of excess charge carriers on the balloon and determine what sign they are. Explain your answer.
(C) The second balloon is now moved closer, to a distance of only 10 cm from the suspended balloon. Describe what happens to the angle the thread makes with the vertical, and explain why this occurs.

132. The electric field around a charged object is shown in the figure.

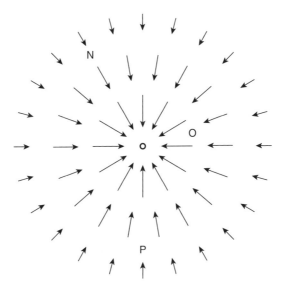

(A) What aspects of the electric field indicate the sign on the charge?
(B) Rank the magnitudes of the electric field at points N, O, and P. Explain what aspects of the diagram indicate the strength of the electric field.
(C) A proton is placed at point P, and an electron is placed at point N. Both are released from rest at the same time. Compare and contrast the acceleration of the two particles at the instant they are released, and explain any differences.
(D) Describe the motions of the proton and the electron for a long time after they are released. Justify your claim.

133. Sketch what the electric field vectors would look like in the space surrounding two closely spaced charges if

   (A) one of the charges is positive and the other is negative (a dipole).
   (B) both charges are positive.

134. Three neutral metal spheres on insulating stands are placed so they touch. A negative rod is brought close to sphere Z, as shown in figure. Then the following sequence of events takes place:
   1. Sphere X is moved away to the left.
   2. The rod is removed.
   3. Finally, sphere Z is moved away from sphere Y.

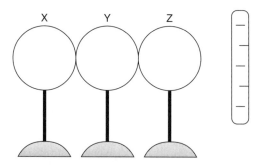

   If the final charge of sphere Z is Q, what are the charges of spheres X and Y? Justify your answer.

135. The metal spheres on insulating stands 1, 2, and 3 are all identical and situated as shown in the figure. Originally sphere 1 and 2 have a charge of $-Q$, and sphere 3 has a charge of $+2Q$. The force on sphere 2 from sphere 1 is $+F$.

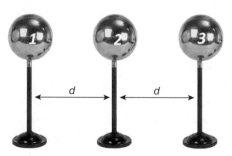

(A) What is the force on sphere 1 from sphere 2 in terms of F?
(B) What is the force on sphere 3 from sphere 1 in terms of F?
(C) Sphere 2 is touched to sphere 1, then to sphere 3, and it is finally replaced in its original position. Rank the magnitude of the final net force on each of the spheres from greatest to least.

136. An electron in a hydrogen atom is 5.29×10^{-11} m from the nucleus.
(A) Calculate the net force on the electron due to the nucleus.
(B) How does the force on the nucleus from the electron compare to your previous answer? Explain why this is the case.
(C) How do the accelerations of the electron and nucleus compare? Explain your answer in terms of orders of magnitude.
(D) Did you need to include the force of gravity between the electron and the nucleus? Explain your answer.

137. A 2-µC and a −8-µC charge are placed as shown on the x-axis.

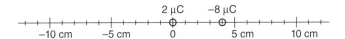

(A) At what location on the x-axis is the electric field equal to zero?
(B) Where would a −4-µC charge be placed so the 2-µC charge would receive a net force of zero?
(C) What charge would be placed at 12 cm so the net force on the −8-µC charge is zero?

Questions 138–142

Two positive charges (+q) are fixed at +d and −d on the y-axis so they cannot move, as shown in the figure.

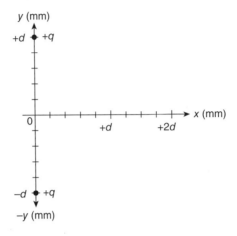

138. Calculate the force on a third charge, −q, placed at +d on the x-axis. What direction is the force? Show all your work.

139. If the charge −q is moved to the origin, what will be the new force on the charge? Justify your response.

140. The charge −q is now moved from the origin along the positive x-axis toward a final location of +2d. During this move, the net electric force on −q changes. Discuss the two conflicting effects of this move, one that will increase and one that will decrease the magnitude of the net electric force.

141. Once at location +2d, the charge −q is released and free to move. Describe the motion of the charge over a long period of time.

142. The −q charge is removed, leaving only the two positive charges +q fixed at +d and −d on the y-axis, as shown in the figure. A third positive charge +q is placed at +d on the x-axis and released. In a clear, coherent, paragraph-length response, describe the motion of the released charge +q over a long period of time. Use the concepts of both forces *and* the energy to explain the motion.

143. The figure shows three different square arrangements of four charges.

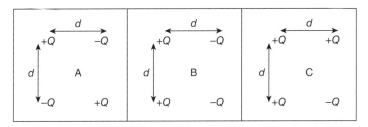

(A) A proton is placed at the center of each square arrangement. Rank the magnitude of the net force on each proton.
(B) Calculate the magnitude and direction of the force on an electron placed at point C.
(C) Calculate the magnitude of the acceleration of the electron placed at point C.

144. Three charges of magnitude q are placed at the corners of an equilateral triangle, as shown in the figure.

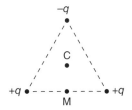

(A) An electron is placed at C, the center of the triangle. Draw a force diagram of all the forces on the electron. All forces should be drawn proportionally. What is the direction of the net force on the electron?
(B) A proton is placed at M, the midpoint of the side of the triangle. Will the net force on the proton be greater than, less than, or the same as the net force on the electron from part (A) above? Justify your claim.

145. A metal sphere on an insulating stand is positively charged, as shown in the figure.

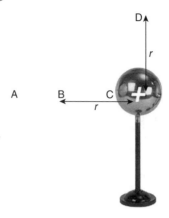

(A) A negatively charged balloon can be placed at location A, B, C, or D. Rank the potential energy of the balloon-metal sphere system when the balloon is placed at each of these locations. Explain your ranking.
(B) A positively charged balloon can be placed at location A, B, C, or D. Rank the potential energy of the balloon-metal sphere system when the balloon is placed at each of these locations. Explain your ranking.

146. Compare and contrast electric and gravitational fields.

Questions 147–150

Two charges, $+Q$ and $-Q$, are placed at the corners of a square whose sides have a length of a. Points P and N are located on the corners of the square. Point O is in the center of the square.

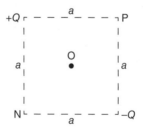

147. Sketch the directions of the electric field at points N, O, and P. Make sure the vectors are drawn to the correct proportion.

148. What are the electric potentials at points N, O, and P?

149. A proton is moved from point P to point O. How much total work is done by the electric field during this move? Explain.

150. By moving only one of the charges, explain how the electric field at point O can be made to point directly to the right.

151. Sketch what the isolines of electric potential look like in the space surrounding two closely spaced charges if
    (A) both charges are negative and have the same magnitude.
    (B) one charge is positive and the other is negative. Both have the same magnitude of charge (a dipole).
    In both cases, indicate the possible numerical values of electric potential for each isoline drawn.

152. Electric field vectors around three charges, 1, 2, and 3, are shown in the figure.

    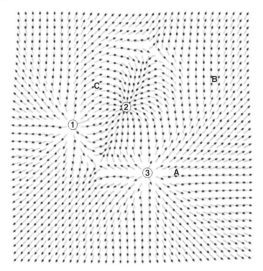

    (A) What are the signs of the three charges? Explain what aspects of the electric field indicate the sign of the charges.
    (B) Draw the direction of the force on an electron placed at point C.
    (C) Sketch two isoline lines of constant electric potential—one that passes through point A and another that passes through point B.
    (D) Which isoline has a higher electric potential, the line that passes through point A or the one that passes through point B? Justify your answer.

Questions 153–159

The figure shows isolines of electric potential. Circles 1 and 2 represent two spherical charges. Points A, B, C, and D represent locations on isolines of electric potential.

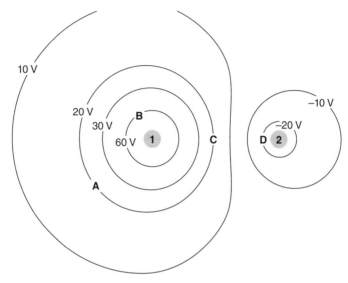

153. What are the signs of the two charges, and how do their relative magnitudes compare? Explain how the isolines help you determine this.

154. The spheres have masses in the same ratio as the magnitude of their charges. Will the isolines of gravitational potential have a similar shape as the isolines shown? Explain.

155. A proton is released from point C and moves through an electric potential difference of magnitude 40 V.
     (A) On which isoline of electric potential will the proton end up?
     (B) The proton will have kinetic energy when it arrives at this new isoline. Where does this kinetic energy come from?
         i. Explain your answer in terms of the system that includes the two charges and the proton.
         ii. Explain your answer in terms of the system that includes only the proton.

156. An electron at point A is moved to point B. Has the electric potential energy of the electron-charges system increased or decreased? Justify your answer with an equation.

157. The distance between points C and D is d. Derive a symbolic expression for the magnitude of the average electric field between the two points. Also, indicate the direction of the electric field.

158. A particle with positive charge of Q is released from point C and gains kinetic energy on its path to point D. Derive a symbolic equation for the amount of work done by the electric field and the final kinetic energy of the proton.

159. Sketch electric field vectors at points A and C. The vectors should be drawn so their relative strengths are reflected in the drawing.

160. Two particles of different charge and mass are separated by a distance of x, as shown in the figure.

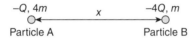

(A) How does the electric force on particle A compare to the electric force on particle B? Explain.
(B) The particles are both released simultaneously. How do the accelerations of the two particles compare? Explain.
(C) Write equations that could be used to solve for the final velocities v_A and v_B of the two particles a long time after they are released in terms of m_A, m_B, Q, and x.

Questions 161–165

The figure shows two metal spheres. Sphere 1 has a radius of R and an initial positive charge of Q_0. Sphere 1 has an initial electric field just outside the surface of E_0 and electric potential at the surface of V_0. Sphere 2 has a radius of $2R$ and is initially uncharged.

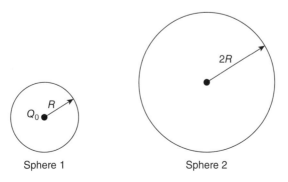

100 › 500 AP Physics 2 Questions to know by test day

161. On the axis, plot the electric field as a function of radius (r), the distance from the center of sphere 1. Briefly explain why the graph has this shape.

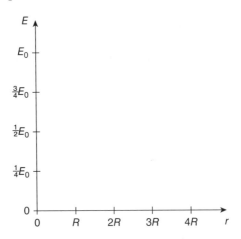

162. On the axis, plot the electric potential as a function of radius (r), the distance from the center of sphere 1. Briefly explain why the graph has this shape.

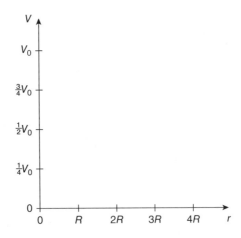

The two spheres are now brought into contact for a few seconds and then placed back in their original positions.

163. How do the final electric potentials of the two spheres (V_1 and V_2) compare after they have touched and been returned to their original positions? Explain.

164. How does the final charge of each sphere Q_1 and Q_2 compare? Explain and write equations that could be used to find the final charge of each sphere.

165. How do the electric fields just outside the surfaces of each sphere E_1 and E_2 compare? Explain your reasoning.

166. A battery of potential difference ΔV is connected to a parallel plate capacitor for a long time. The separation between the plates is d, and the area of one plate is A.

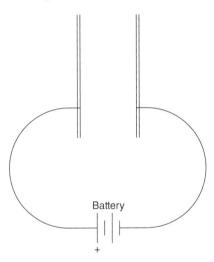

(A) Sketch the electric field between the plates of the capacitor.
(B) Sketch isolines of constant electric potential between the plates.
(C) Write an expression for the electric field strength between the plates.
(D) Write an expression for the charge on the left plate. Show your work.
(E) What is the net charge on both plates combined? Explain.
(F) A proton with a charge of $+e$ is released from the positive plate. Write an expression for the net force on the proton using known quantities. Do you need to include the force of gravity in your calculation? Justify your answer.

(G) Write an expression for the velocity of the proton when it reaches the negative plate. Derive this value using the concept of forces and the concept of energy.

(H) Now a second proton is released from a point midway between the plates. Does this proton reach the negative plate with the same velocity as the first proton that was released from the positive plate? Justify your answer with an equation.

167. A charged parallel plate capacitor with a distance of *d* between the plates is shown in the figure. A point in the middle of the capacitor is marked with a dot. The dot is also located inside a dashed square with locations marked along the sides and corners. Eight paths are shown emanating from the dot to each marked location.

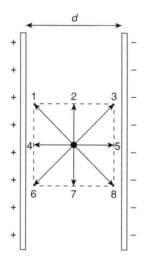

(A) Along which path would you move an electron to increase its electric potential energy? Explain your answer.

(B) From which of the marked locations would you release a proton so it would attain the greatest velocity after release? Justify your answer.

(C) How do the electric fields at each of the marked locations compare?

Questions 168 and 169

A proton is launched between the plates of a charged parallel plate capacitor, as shown in the figure.

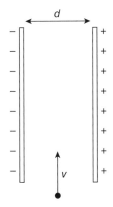

168. Discuss the similarities and differences between the motion of the proton and that of a baseball thrown horizontally in the Earth's gravitational field.

169. The proton is replaced with an electron launched with the same velocity. Discuss how the motion of the electron differs from that of the proton.

Questions 170–174

A parallel plate capacitor with a capacitance of C is shown in the figure. The area of one plate is A, and the distance between the plates is d.

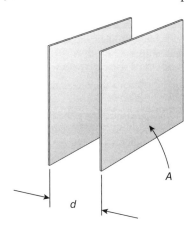

170. If the area of both capacitor plates as well as the distance between them were doubled, what would be the effect on the capacitance of the capacitor? Explain.

171. The capacitor is connected to a battery of potential difference ΔV. If the potential difference of the battery is doubled, what happens to the charge stored on the plates and the capacitance of the capacitor? Justify your answer.

172. In an experiment, the area (A) of the capacitor plates is changed to investigate the effect on the capacitance (C) of the capacitor. Sketch the graph of the lab data you expect to see from this experiment.

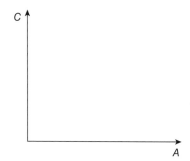

173. In another experiment, the distance between the plates (d) is changed to investigate the effect on the capacitance (C) of the capacitor. Sketch the graph of the lab data you expect to see from this experiment.

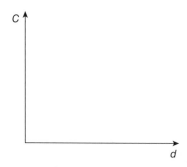

174. You are going to use a capacitor to power a lightbulb. You need the bulb to shine for a long time. Describe the geometry of the capacitor you would choose to power the bulb. Explain your answer.

Questions 175 and 176

A charged capacitor generates the electric field shown in the figure.

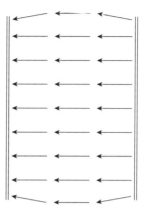

175. An electric dipole shown below is placed between the plates of the capacitor. Draw any electric forces experienced by the dipole, and describe the initial motion of the dipole.

176. An uncharged metal box is placed between the plates of the capacitor.

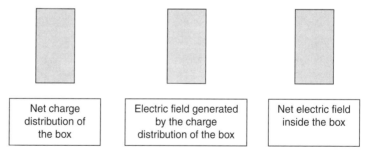

| Net charge distribution of the box | Electric field generated by the charge distribution of the box | Net electric field inside the box |

(A) Draw any net charge on the box that results from it being placed inside the capacitor on the figure on the left.
(B) Draw any electric field generated by the charges of the box on the figure in the middle.
(C) Draw the net electric field inside the box due to its own charge distribution and due to the external electric field of the capacitor on the figure on the right.

AP-Style Multiple-Choice Questions

177. A positively charged rod is brought near to but not touching three metal spheres that are in contact with each other, as shown in the figure. Which is the best representation of the charge arrangement inside the three spheres?

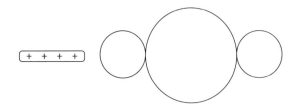

(A)

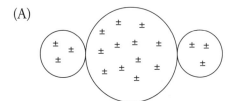

(B)

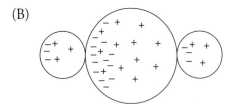

(C)

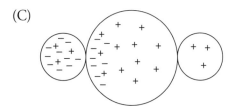

(D)

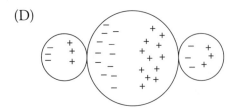

178. Isolines of equal electric potential in a region of space are shown in the figure. Points A and B are in the plane of the isolines. Which of the following correctly describes the relationship between the magnitudes and directions of the electric fields at points A and B?

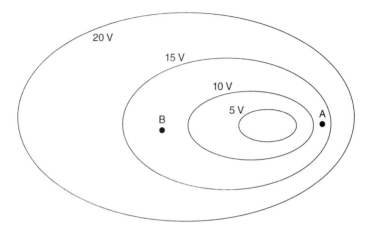

(A) $E_A = E_B$ and is in the same direction.
(B) $E_A \neq E_B$ and is in the same direction.
(C) $E_A = E_B$ and is in the opposite direction.
(D) $E_A \neq E_B$ and is in the opposite direction.

179. Two electrons exert an electrostatic repulsive force on each other. Is it possible to arrange the two electrons so the gravitational attraction between them is large enough to cancel out the electric repulsive force?

(A) No, the charge of the electrons squared is much larger than the mass of the electrons squared.
(B) No, there is no gravitational force between subatomic particles.
(C) Yes, reducing the radius between the electrons will increase the gravitational force as it is proportional to the inverse of the radius squared.
(D) Yes, increasing the distance between the electrons will reduce the electrostatic repulsion until it is equal to the gravitational force.

180. Two charges (−2Q and +Q) are located as shown in the figure. Three regions are designated in the figure: X is to the left of −2Q; Y is between the two charges; and Z is to the right of +Q. Which of the following correctly ranks the magnitude of electric field in the three regions?

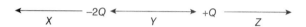

(A) $E_X > E_Y > E_Z$
(B) $E_Y > E_X > E_Z$
(C) $E_Y > E_X = E_Z$
(D) It is not possible to rank the magnitudes of the electric fields without more information.

181. A balloon that has been rubbed with hair is suspended from the ceiling by a light thread. One at a time, a neutral wooden board and then a neutral steel plate of the same size and shape are brought near to the balloon without touching. Which of the following correctly describes and explains the behavior of the balloon?

(A) The balloon is not attracted to the steel or the wood because both are neutral objects.
(B) The balloon is attracted to the steel because it is a conductor but not to the wood because it is an insulator.
(C) The balloon is attracted to both the steel and wood equally because both become polarized.
(D) The balloon is attracted to the steel more than it is attracted to the wood because the steel polarizes with a larger charge separation.

182. A negatively charged metal rod is brought close to a neutral metal sphere without touching. Which of the following is correct concerning the final state of the metal sphere?

(A) The sphere contains no changes because the rod does not touch the sphere.
(B) The sphere acquires a net positive charge by induction.
(C) The sphere acquires a net negative charge by conduction.
(D) The sphere remains neutral, but the charge distribution changes due to polarization.

183. The news reports the discovery of two new particles by the research facility CERN in Geneva. The first particle, dubbed Alithísium, is large with a mass equivalence of 125 GeV/c^2 ± 15 GeV/c^2 and a net charge of -1.55×10^{-18} C ± 0.1×10^{-18}. The second particle, Psevdísium, has a mass of only 5.4×10^{-4} u ± 0.1×10^{-4} u and a charge of 1.6×10^{-20} C ± 0.5×10^{-20}. Which of the following is most correct concerning the two new particles?

   (A) Both particles appear reasonable.
   (B) Alithísium appears reasonable, but Psevdísium does not.
   (C) Psevdísium appears reasonable, but Alithísium does not.
   (D) Neither particle appears reasonable.

184. Two neutral metal spheres on insulating stands are placed so they touch, as shown in figure a. A positive rod is brought close to sphere A, as shown in figure b. Sphere B is moved to the right, as shown in figure c. The positive rod is then removed, as shown in figure d. Which of the following correctly describes the situation after the rod is removed? *Select two answers.*

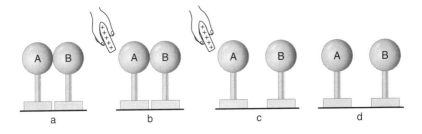

   (A) The net charge of the system that includes both spheres remains neutral.
   (B) The net charge of sphere B is negative.
   (C) Spheres A and B attract each other.
   (D) The electric field between the spheres points to the right.

185. The metal spheres on insulating stands 1, 2, and 3 are all identical and situated as shown in the figure. Spheres 1 and 2 have a charge of $-Q$, and sphere 3 has a charge of $+2Q$. The force of sphere 1 on sphere 2 is $+F$. What is the magnitude of the net force on sphere 3 in terms of F?

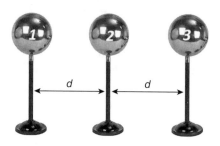

(A) $\frac{3}{2}F$

(B) $2F$

(C) $\frac{5}{2}F$

(D) $3F$

186. A positive charge $(+q)$ is placed at vertex A of a triangle, as shown in the diagram. What charge must be placed at vertex B to cause an electron placed at vertex C to receive a force as shown?

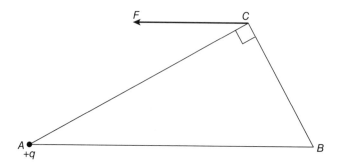

(A) Positive and smaller than $|+q|$
(B) Positive and larger than $|+q|$
(C) Negative and smaller than $|+q|$
(D) Negative and larger than $|+q|$

187. The figure shows isolines of electric potential in a region of space. Which of the following will produce the greatest increase in electric potential energy of the particle in the electric field?

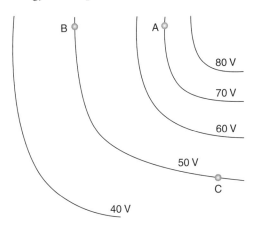

(A) Moving an electron from point A to point C
(B) Moving an electron from point B to point A
(C) Moving a proton from point B to point C
(D) Moving a proton from point A to point C

188. The figure shows isolines of constant electric potential surrounding two charges. Which of the following correctly describes the two charges?

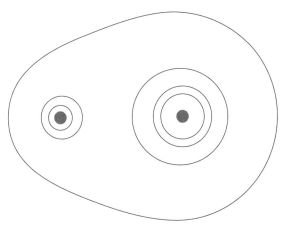

(A) The charges are the same magnitude and the same sign.
(B) The charges are the same magnitude but different signs.
(C) The charges are different magnitudes but the same sign.
(D) The charges are different magnitudes and different signs.

Questions 189–193

The left figure shows a capacitor with a horizontal electric field. The distance between the plates is $4x$. The right figure shows two electrons, e_1 and e_2, and two protons, p_1 and p_2, which are placed between the plates at the locations shown.

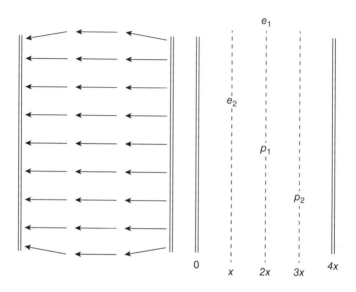

189. Which of the following correctly ranks the electric fields at locations x, $2x$, and $3x$?
    (A) $E_x > E_{2x} > E_{3x}$
    (B) $E_x = E_{2x} = E_{3x}$
    (C) $E_x = E_{3x} > E_{2x}$
    (D) $E_{3x} > E_{2x} > E_x$

190. Which of the following is a correct statement about the forces on the charges?
    (A) The forces on e_1 and e_2 are not the same in magnitude but are the same in direction.
    (B) All four particles receive the same magnitude of force but not all in the same direction.
    (C) The force on p_1 is the largest in magnitude because it is in the middle of the capacitor where the electric field is strongest.
    (D) The forces on e_2 and p_2 are the largest in magnitude because they are closer to the charged plates.

191. Charge p_1 is released from rest. Which of the trajectories shown in the figure is a possible path of the released charge?

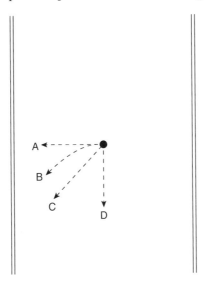

(A) A
(B) B
(C) C
(D) D

192. All the particles are released from rest from the locations shown. Which particle will achieve the greatest magnitude of velocity?

(A) e_1
(B) e_2
(C) p_1
(D) p_2

193. After being released from rest, proton p_2 attains a final velocity of v just before striking a capacitor plate. Let the mass and charge of the proton be m_p and e. The electric potentials at locations 0, x, $2x$, $3x$, and $4x$ are V_0, V_x, V_{2x}, V_{3x}, and V_{4x}, respectively. What is the magnitude of the electric field between the plates? *Select two answers.*

(A) $\dfrac{V_{4x} - V_{3x}}{3x}$

(B) $\dfrac{V_0 - V_{3x}}{3x}$

(C) $\dfrac{m_p v^2}{6xe}$

(D) $\dfrac{e}{36\pi\varepsilon_0 x^2}$

AP-Style Free-Response Questions

194. Two charges ($-2q$ and $+q$) are situated along the x-axis as shown in the figure.

(a) Derive an expression for the electric field at $-2x$. Show all your work.
(b) If a charge of $5q$ is placed at $-2x$, what will be the magnitude and direction of the force on $5q$? Show all your work.
(c) What is the direction of the electric field at point $+2x$?
(d) Rank the electric fields' strength at points $-2x$, 0, and $+2x$.
(e) On the axis, sketch a graph of the electric field E along the x-axis as a function of position x. Electric fields to the right are defined as positive.

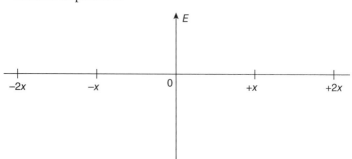

(f) Is there a spot on the *x*-axis where the electric field will have a magnitude of zero? If so, give the general location of where it will be. If not, explain why not.

195. Three positive charges (+*q*) are fixed at the vertices of an equilateral triangle, as shown in the figure. Point C is at the center of the triangle. Point M is the midpoint of the bottom leg of the triangle. The distance between C and M is *x*.

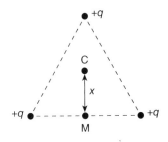

(a) At which point, C or M, is the electric field stronger? Justify your answer.
(b) Select a proton or electron to place at either location C or M such that the particle will receive an upward electric force (toward the top of the page). Explain your reasoning.
(c) A particle is moved from point M to point C.
  i. Derive an expression for the change in the magnitude of the electric potential energy of the four-charge system in terms of the magnitude of the charge (Q) and the electric potentials at points M and C (V_M, V_C).
  ii. Derive a symbolic expression for the magnitude of work done on the charge by the electric field in terms of $E_{average}$, the average electric field between points M and C.
(d) The topmost of the three +*q* charges is released and accelerates upward and away. Will the released charge continue to accelerate indefinitely? Yes or no.
  i. Explain your answer in terms of forces.
  ii. Explain your answer in terms of energy.

196. A scientist is attempting to determine the velocity of electrons ejected from the cathode of an old TV set. The scientist builds the apparatus shown in the figure, which consists of two oppositely charged parallel conducting plates, each with an area (A) of 0.300 m², separated by a distance (d) of 0.020 m. Each plate has a hole through which the electrons from the cathode are injected into the apparatus. Electrons enter through the left hole and exit through the right hole. A sensor is placed on the right side of the apparatus to count electrons exiting the right plate. The scientist connects a variable voltage battery to the apparatus and increases the voltage until no more electrons exit the right hole. No electrons exit the apparatus when the electric potential difference between the plates is $\Delta V = 500$ V.

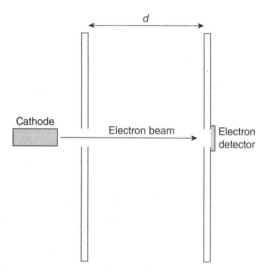

(a) Calculate the magnitude and direction of the electric field between the plates when no electrons exit the apparatus on the right-hand side. Explain the direction of the electric field by referencing the electric force on the particle and acceleration of the particle.
(b) Calculate the speed of the fastest electrons emitted by the cathode.
(c) How long does it take the fastest electron to cross the distance between the plates?
(d) Calculate the magnitude of the charge on each plate.

The apparatus is now turned on its side so the electron beam can be shot down the middle of the two plates, as shown in the figure. The electric potential difference between the plates is again 500 V. The electron detector is moved along the top plate until it locates where the electrons are striking the upper plate at a distance of *x* from the entry point of the electrons.

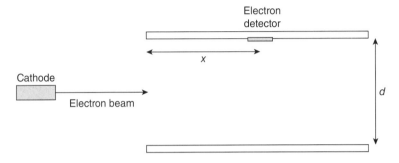

(e) Sketch the path of the electrons to the detector.
(f) On the axis, sketch the *x* and *y* velocities of the electrons from the point where they enter the region between the plates until they strike the detector. The positive *x-y* directions are to the right and upward, respectively.

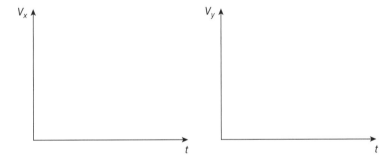

(g) The cathode is replaced with a radioactive material that emits alpha particles with the same velocity and direction as the electrons emitted by the cathode. In a clear, concise, paragraph-length response, explain where a detector would be placed to locate the impact of the alpha particles on the plate of the apparatus.

Electric Circuits

Skill-Building Questions

197. Explain what *electric current* means physically, and explain the difference between conventional current and electron current.

198. Explain the distinctions between electric potential, electric potential difference, emf, and voltage.

Questions 199–202

Two circuits are shown in the figure. Circuit #1 has two identical resistors. Circuit #2 has only one resistor. Assume the electric potential of point A is zero and that the wires in the circuit have a small but measurable resistance.

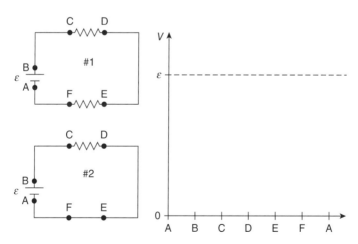

199. Using a dashed line, sketch the electric potential of circuit #1 starting at point A and moving clockwise around the circuit back to point A.

200. How would the graph for circuit #1 change if the resistor between points C and D were increased to twice the size of the resistor between points E and F? Justify your answer.

201. How would the sketch be different if the resistance of the wires were negligible? Explain.

202. Using a solid line, sketch the electric potential of circuit #2 starting at point A and moving clockwise around the circuit back to point A. Make sure your sketch is in proportion to the previous graph.

Questions 203–206

Two circuits are shown in the figure. Circuit #1 has two identical resistors. Circuit #2 has only one resistor. Assume the electric potential of point A is zero and that the wires in the circuit have a small but measurable resistance.

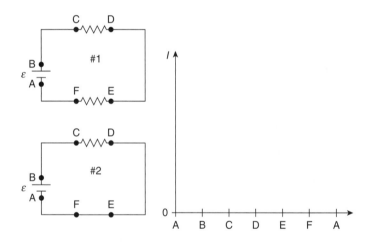

203. Using a dashed line, sketch the current of circuit #1 starting at point A and moving clockwise around the circuit back to point A.

204. How would the graph for circuit #1 change if the resistor between points C and D were increased to twice the size of the resistor between points E and F? Justify your answer.

205. How would the sketch be different if the resistance of the wires were negligible? Explain.

206. Using a solid line, sketch the current of circuit #2 starting at point A and moving clockwise around the circuit back to point A. Make sure your sketch is in proportion to the previous graph.

Questions 207–212

A circuit with two identical resistors has two pathways, as shown in the figure. Assume the electric potential of point A is zero and that the wires in the circuit have a small but measurable resistance.

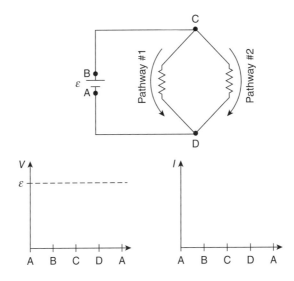

207. On the potential graph, use a dashed line to sketch the electric potential of the circuit starting at point A and moving clockwise around the circuit through pathway #1 and back to point A.

208. How would the electric potential graph for pathway #1 change if the resistor between points C and D were increased to twice its current size? Justify your answer.

209. On the potential graph, use a solid line to sketch the electric potential of the circuit starting at point A and moving clockwise around the circuit through pathway #2 and back to point A. Make sure your sketch is in proportion to the previous graph.

210. On the current graph, use a dashed line to sketch the current of the circuit starting at point A and moving clockwise around the circuit through pathway #1 and back to point A.

211. How would the graph of the current through pathway #1 change if the resistor in pathway #1 were increased to twice its current size? Justify your answer.

212. On the current graph, use a solid line to sketch the current of the circuit starting at point A and moving clockwise around the circuit through pathway #2 and back to point A. Make sure your sketch is in proportion to the previous graph.

213. In the following table, indicate to what point or points you would connect a voltmeter or ammeter to measure the potential difference and current for the listed circuit element. Assume the wires have negligible resistance.

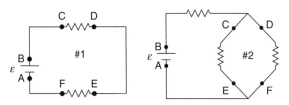

Circuit element	Point or pair of points to which you would connect a voltmeter to measure the potential difference across the circuit element	Point or pair of points to which you would connect an ammeter to measure the current through the circuit element
Battery in circuit #1		
Top resistor in circuit #1		
Battery in circuit #2		
Far right resistor in circuit #2		

214. Explain the difference between ohmic and non-ohmic objects.

215. How would you discover whether or not a material is ohmic?

216. A resistor has a length of L, a cross-sectional area of A, and a diameter of D, as shown in the figure.

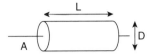

   (A) What happens to the resistance of a resistor when its length is doubled?
   (B) What happens to the resistance of a resistor when its cross-sectional area is tripled?
   (C) What happens to the resistance of a resistor when its diameter is quadrupled?

217. Explain Kirchhoff's loop rule, how it is used, and what kind of conservation it represents.

218. Explain Kirchhoff's junction rule, how it is used, and what kind of conservation it represents.

219. A circuit with four resistors of different resistances (R_1, R_2, R_3, and R_4), a battery of potential difference (ε), and an ammeter (A_1) is shown in the figure.

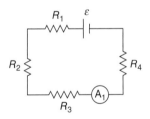

   (A) Explain why the current is the same through all the resistors in the circuit without using the word *series*. (Hint: Use one of Kirchhoff's laws.)
   (B) Write Kirchhoff's loop rule for this circuit.
   (C) Write the expression for the equivalent resistance R_{eq} of the circuit in terms of known quantities.
   (D) Write the expression for the current I_1 measured by the ammeter in terms of known quantities.
   (E) Write an expression for the potential difference ΔV_2 across resistor R_2 in terms of known quantities.

220. A circuit with two resistors of resistance (R_1 and R_2), a battery of potential difference (ε), and three ammeters (A_1, A_2, and A_3) is shown in the figure.

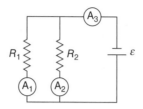

(A) Explain why the potential difference is the same across both resistors in the circuit without using the word *parallel*. (Hint: Use one of Kirchhoff's laws.)
(B) Write Kirchhoff's loop rule for the loop that contains the battery and resistor R_1 and for the loop that contains the battery and resistor R_2.
(C) Write Kirchhoff's junction rule for this circuit.
(D) Write an expression for the equivalent resistance R_{eq} of the circuit in terms of known quantities.
(E) Write a separate expression for the currents I_1, I_2, and I_3 measured by the three ammeters in terms of known quantities.

221. A circuit with four identical resistors (R_1, R_2, R_3, and R_4), a battery of potential difference (ε), and four ammeters (A_1, A_2, A_3, and A_4) is shown in the figure.

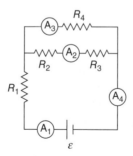

(A) Use Kirchhoff's junction rule to prove that currents I_1 and I_4 are the same.
(B) Write Kirchhoff's loop rule for the loop that contains the battery and resistor R_3.
(C) Rank the currents (I_1, I_2, I_3, and I_4) from greatest to least. Justify your answer.
(D) Write an expression for the equivalent resistance of the circuit in terms of known quantities.

222. Two lightbulbs have power ratings of 40 W and 100 W when connected to a potential difference of 120 V.
   (A) Calculate the resistance of both bulbs. Show your work.
   (B) Which bulb glows brightest when connected in parallel? Justify your answer.
   (C) Which bulb glows brightest when connected in series? Justify your answer.

Questions 223–226

A circuit consisting of a battery of potential difference (ε) and with internal resistance (r) is connected to a switch (S) and resistor (R), as shown in the figure.

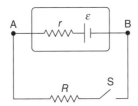

223. Write Kirchhoff's loop rule for this circuit.

224. Use your Kirchhoff's loop rule equation from above to write an expression for the internal resistance of the battery. Based on this equation, what information would you need to determine the emf and internal resistance of the battery?

225. Use your Kirchhoff's loop rule equation to derive an equation for the potential difference across the battery V_{AB}. (This is also called the terminal voltage of the battery.) How can we use this equation to find the emf of the battery?

226. On the axis provided, sketch the graph of terminal voltage of the battery as a function of current through the battery. Explain the significance of both intercepts and the slope of the line.

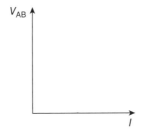

227. Three identical bulbs are attached to batteries, as shown in the figure. Rank the brightness of the bulbs. Justify your prediction in terms of power.

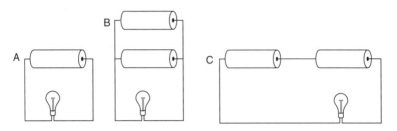

228. Four identical bulbs are attached in a circuit, as shown in the figure. Rank the brightness of the bulbs. Justify your prediction in terms of power.

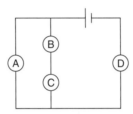

229. The graph shows the power supplied by a battery in a circuit where the resistance is changed. Use the slope of the graph to find the electric potential of the battery. Explain how this was done.

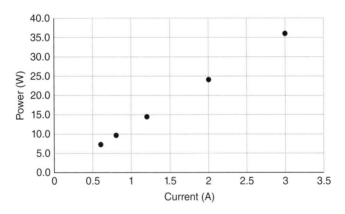

230. The graph shows power dissipated by a resistor when connected to varying voltage. Use the slope of the graph to calculate the resistance of the resistor. Show your work.

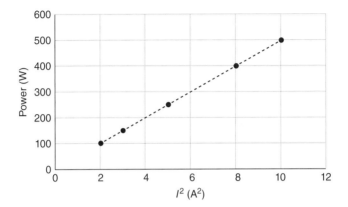

231. The graph shows power dissipated by a resistor when connected to varying voltage. Use the slope of the graph to calculate the resistance of the resistor. Show your work.

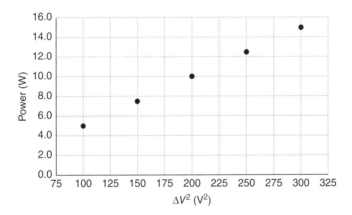

232. Using standard electrical schematic figures, sketch a circuit with bulbs, a battery, and switches that will accomplish the following.
   (A) Two bulbs with a switch that will turn all the bulbs on/off
   (B) Three bulbs with a switch that will turn only one bulb on/off while the others stay on
   (C) Three bulbs with switches that will turn each bulb on/off individually and a master switch that will turn all the bulbs on/off

233. Using standard electrical schematic figures, sketch a circuit with bulbs, a battery, a capacitor, and switches that will accomplish the following. Assume that the capacitor is initially uncharged.
(A) When the switch is closed, the bulb will immediately light but over time will go out.
(B) When the switch is closed, the bulbs will not immediately light, but over time they will glow.

234. Complete the voltage-current-resistance-power (VIRP) chart for the circuit shown in the figure.

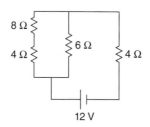

Component	V	I	R	P
R_1			4 Ω (on the right)	
R_2			6 Ω	
R_3			4 Ω (on the left)	
R_4			8 Ω	
Total for circuit	12 V			

235. Complete the voltage-current-resistance-power (VIRP) chart for the circuit shown in the figure. The emf of the battery is unknown. The ammeter measures a current of 1.82 A.

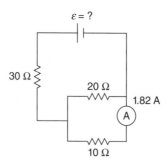

Component	V	I	R	P
R_1			30 Ω	
R_2			20 Ω	
R_3			10 Ω	
Total for circuit				

236. Complete the voltage-current-resistance-power (VIRP) chart for the circuit shown in the figure. One resistor has an unknown resistance. The ammeter measures 0.67 A.

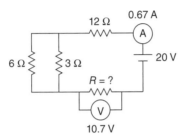

Component	V	I	R	P
R_1			6 Ω	
R_2			3 Ω	
R_3			12 Ω	
R_4			?	
Total for circuit	20 V			

237. The circuit in the figure has three identical resistors, two ammeters, a battery, and a switch. The switch is originally open. When the switch is closed, what happens to the reading in each ammeter? Justify your answer.

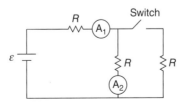

238. The circuit in the figure consists of two capacitors (2 µF and 4 µF) connected to a 200 V battery.

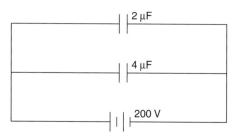

(A) Calculate the equivalent capacitance of the two capacitors.
(B) Calculate the energy stored in the 4-µF capacitor.

239. The circuit in the figure consists of three capacitors (3 µF, 4 µF, and 6 µF) connected to a 200 V battery.

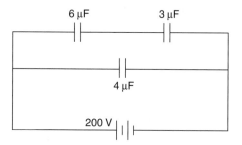

(A) Calculate the equivalent capacitance of the combined three capacitors.
(B) Calculate the total energy stored in the 6-µF and 3-µF capacitor combination.

240. The circuit shown in the figure consists of three identical resistors, two ammeters, a battery, a capacitor, and a switch. The capacitor is initially uncharged, and the switch is open. Explain what happens to the readings of the two ammeters from the instant the switch is closed until a long time has passed.

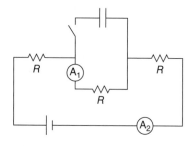

241. The figure shows a circuit with three resistors, a battery, a capacitor, a switch, and three ammeters. Originally the switch is open, and the capacitor is uncharged.

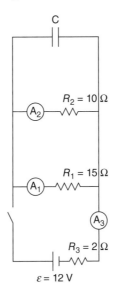

(A) Complete the voltage-current-resistance-power (VIRP) chart for the circuit immediately after the switch is closed.

Location	V	I	R	P
1			15 Ω	
2			10 Ω	
3			2.0 Ω	
Total for circuit	12 V			

(B) Complete the voltage-current-resistance-power (VIRP) chart for the circuit after the switch is closed for a long time.

Location	V	I	R	P
1			15 Ω	
2			10 Ω	
3			2.0 Ω	
Total for circuit	12 V			

242. The figure shows a circuit with two resistors, a battery, a capacitor, and a switch. Originally, the switch is open, and the capacitor is uncharged.

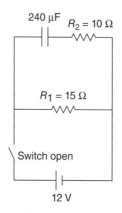

(A) Complete the voltage-current-resistance-power (VIRP) chart for the circuit immediately after the switch is closed.

Location	V	I	R	P
1			15 Ω	
2			10 Ω	
Total for circuit	12 V			

(B) Complete the voltage-current-resistance-power (VIRP) chart for the circuit after the switch is closed for a long time.

Location	V	I	R	P
1			15 Ω	
2			10 Ω	
Total for circuit	12 V			

(C) What is the energy stored in the capacitor after the switch has been closed a long time?

Electric Circuits 133

AP-Style Multiple-Choice Questions

Questions 243 and 244

Three cylindrical resistors made of the same material but different dimensions are connected, as shown in the figure. A battery is connected across the resistors to produce current.

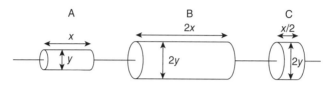

243. Which is the correct ranking of the currents for the resistors?
 (A) $I_A = I_B = I_C$
 (B) $I_A > I_B > I_C$
 (C) $I_C > I_A = I_B$
 (D) $I_C > I_B > I_A$

244. Which is the correct ranking of the potential differences of the resistors?
 (A) $V_A = V_B = V_C$
 (B) $V_A > V_B > V_C$
 (C) $V_A = V_B > V_C$
 (D) $V_C > V_B > V_A$

Questions 245–249

The figure shows current as a function of electric potential difference for a resistor and bulb.

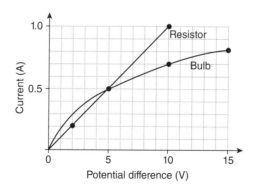

245. Are the devices ohmic?

	Resistor	Bulb
(A)	Ohmic	Ohmic
(B)	Ohmic	Non-ohmic
(C)	Non-ohmic	Ohmic
(D)	Non-ohmic	Non-ohmic

246. If the resistor and bulb are connected in parallel to a 10.0 V battery, what is the total current passing through the system?

(A) 0.5 A
(B) 0.7 A
(C) 1.0 A
(D) 1.7 A

247. With the resistor and bulb still connected in parallel to the 10.0 V battery, what is the total power dissipated by the bulb and resistor?

(A) 0.042 W
(B) 10 W
(C) 17 W
(D) 24 W

248. What is the equivalent resistance of the bulb and resistor while still connected in parallel to the 10.0 V battery?

(A) 0.17 Ω
(B) 5.9 Ω
(C) 17 Ω
(D) 24 Ω

249. The bulb and resistor are removed and reconnected in series to the 10.0 V battery. What is the total current passing through the system?

(A) 0.41 A
(B) 0.50 A
(C) 1.0 A
(D) 1.7 A

Questions 250–251

Two batteries and two resistors are connected in a circuit, as shown in the figure. The currents through R_1, R_2, and ε_2 are shown.

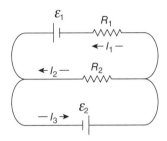

250. Which of the following is a proper application of conservation laws to this circuit? *Select two answers.*
    - (A) $\varepsilon_2 - I_2 R_2 = 0$
    - (B) $\varepsilon_1 - \varepsilon_2 - I_1 R_1 = 0$
    - (C) $I_1 + I_2 - I_3 = 0$
    - (D) $I_2 + I_3 - I_1 = 0$

251. The resistors R_1 and R_2 have the same resistance. If the potential differences of the batteries are $\varepsilon_1 = 9$ V and $\varepsilon_2 = 6$ V, which resistor will have the most current passing through it?
    - (A) R_1
    - (B) R_2
    - (C) R_1 and R_2 have the same current.
    - (D) It is not possible to determine the currents through the resistors without more information.

252. The figure shows two bulbs connected to a battery in a circuit with a switch that is originally in the closed position. What happens to the brightness of the bulbs when the switch is opened?

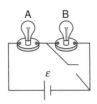

    Bulb A   Bulb B
(A) Four times brighter   Goes out
(B) Same brightness as originally   Glows as brightly as bulb A
(C) Half as bright as originally   Glows as brightly as bulb A
(D) Quarter as bright as originally   Glows as brightly as bulb A

Questions 253 and 254

Four identical resistors of resistance R are connected to a battery, as shown in the figure. Ammeters A_1 and A_2 measure currents of 1.2 A and 0.4 A, respectively.

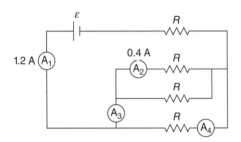

253. What are the currents measured by ammeters A_3 and A_4?

        A_3 A_4
(A) 0.4 A   0.4 A
(B) 0.8 A   0.4 A
(C) 0.4 A   1.2 A
(D) 0.8 A   1.2 A

254. What is the equivalent resistance of the circuit?

(A) $\frac{1}{4}R$

(B) $\frac{4}{3}R$

(C) $\frac{5}{2}R$

(D) $4R$

255. Two resistors made of the same material are shown in the figure. A current of I flows through the left resistor when connected to a potential difference of V. What current will flow through the right resistor when connected to the same potential?

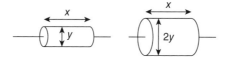

(A) $\frac{I}{2}$

(B) I

(C) $2I$

(D) $4I$

256. A student is given a battery with an unknown emf (ε) and an internal resistance of r. The student sets up a circuit with a known resistor and switch, as shown in the figure. Which measurements should the student make to find the values of both ε and r? *Select two answers.*

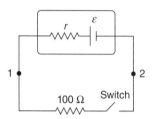

(A) With the switch open, measure the potential difference between points 1 and 2 and the current at point 1.
(B) With the switch closed, measure the potential difference between points 1 and 2 and the current at point 1.
(C) With the switch open, measure the potential difference between points 1 and 2. Close the switch and measure the current at point 1.
(D) With the switch open, measure the potential difference between points 1 and 2. Close the switch and measure the potential difference between points 1 and 2.

257. The circuit shown in the figure has two resistors, an uncharged capacitor, a battery, two ammeters, and a switch initially in the open position. What will happen to the current measured in the ammeters from the instant the switch is closed to a long time after the switch is closed?

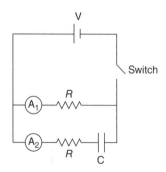

	Ammeter 1	Ammeter 2
(A)	Reading remains constant	Reading remains constant
(B)	Reading remains constant	Reading will change
(C)	Reading will change	Reading remains constant
(D)	Reading will change	Reading will change

258. Four identical capacitors with a plate area of A, a distance between the plates of d, and a dielectric constant κ are connected to a battery, a resistor, and a switch in series. The switch is closed for a long time. The total energy stored in the set of four capacitors is U. The four capacitors in series are to be replaced with a single capacitor that will store the same energy as the four-capacitor set. Which capacitor geometry will accomplish this? *Select two answers.*

	Dielectric constant	Plate area	Distance between plates
(A)	2κ	$2A$	d
(B)	κ	$2A$	$2d$
(C)	$\frac{1}{2}\kappa$	$\frac{1}{2}A$	d
(D)	κ	A	$4d$

Questions 259 and 260

A single resistor is connected to a voltage source that consists of batteries with the same voltage connected in series. The power dissipated by the resistor for various voltages is shown in the two graphs.

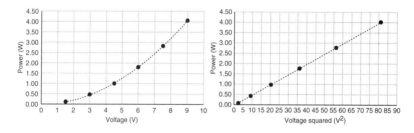

259. Which of the following can be deduced from the graphs? *Select two answers.*
(A) The batteries have a potential difference of 1.5 V.
(B) The batteries have internal resistance.
(C) The resistor is non-ohmic.
(D) The power dissipated by the resistor is proportional to the voltage squared.

260. What is the resistance of the resistor?
(A) 0.05 Ω
(B) 0.22 Ω
(C) 4.5 Ω
(D) 20 Ω

Questions 261 and 262

A resistor with a resistance of R is sealed in a closed cylindrical container. The gas inside the cylinder has an initial pressure of P, an initial volume of V, and N number of atoms. A battery of emf (ε) is connected to the resistor through an airtight piston of mass m fitted inside the cylinder. The piston is able to move up and down as energy is supplied to the gas by the electrical circuit.

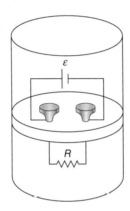

261. Which of the following is an expression of the change in temperature ΔT of the gas while energy is supplied to the gas by the resistor?

(A) $\Delta T = \dfrac{2\varepsilon^2}{3Nk_b R}$

(B) $\Delta T = \dfrac{2\varepsilon^2 t}{3Nk_b R}$

(C) $\Delta T = \dfrac{2\varepsilon t}{3Nk_b R}$

(D) $\Delta T = \dfrac{2\varepsilon}{3Nk_b R}$

262. The temperature of the gas, as a function of time for the single resistor circuit, is shown in the figure. The apparatus is now opened, and a second identical resistor R is connected in parallel to the circuit inside the cylinder. The apparatus is then resealed and reset to the original conditions. How will this affect the temperature versus time graph?

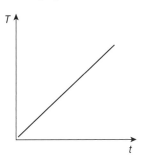

(A) The slope of the graph quadruples.
(B) The slope of the graph doubles.
(C) The slope of the graph is cut in half.
(D) The slope of the graph is one-quarter as large.

AP-Style Free-Response Questions

263. Some students are investigating how the geometry of the cylindrical shaft of graphite in the center of a pencil influences the resistance of the conducting pathway. The students use a 9 V battery as an emf source.

    In the first part of the investigation, the students choose to investigate the influence of length on the resistance of the graphite conductive pathway.

    (a) i. Besides the graphite and battery, what additional equipment would you need to gather the data needed to determine the influence of length on the resistance of the graphite conductive pathway?
        ii. Using standard symbols for circuit elements, draw a schematic diagram of the circuit the students could use to determine the influence of length on the resistance of the conductive pathway. Include the appropriate locations and electrical connection of all equipment including any measuring devices. Clearly label your diagram.

iii. Describe the procedure you would use with your circuit to gather enough data to determine the influence of length on the resistance of the conductive pathway. Make sure your procedure is detailed enough that another student could perform the experiment.

iv. The 9 V battery used in the experiment has a sizable internal resistance. Would you need to change your procedure in part iii? Justify your answer.

Next the students investigate how the geometry of Play-Doh influences the resistance of cylindrical lengths of Play-Doh used as a conductive pathway. The investigation results in the data in the table.

Trial	Diameter (m)	Length (m)	Current (A)	Voltage across Play-Doh (V)			
1	0.002	0.1	0.003	9.0			
2	0.002	0.2	0.001	9.0			
3	0.002	0.3	0.001	9.0			
4	0.002	0.4	0.001	9.0			
5	0.002	0.5	0.001	9.0			
6	0.003	0.1	0.006	9.0			
7	0.003	0.2	0.003	9.0			
8	0.003	0.3	0.002	9.0			
9	0.004	0.1	0.011	9.0			
10	0.004	0.2	0.006	9.0			
11	0.006	0.1	0.025	8.9			
12	0.008	0.1	0.045	8.9			
13	0.010	0.1	0.069	8.8			

(b) i. Which subset of data would be most useful in creating a graph to determine the relationship between the resistance and diameter of the Play-Doh? If the data chosen are incomplete, fill in the needed data in the extra columns provided in the table.

ii. Plot the subset of data you chose on the axis, being sure to label the axis. Draw a line or curve that best represents the relationship between the variables.

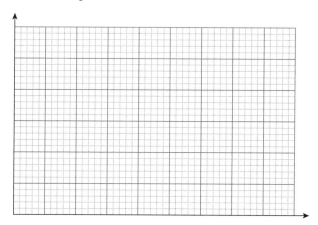

iii. What can you conclude from your line or curve about the relationship between the resistance and diameter of the conductive pathway?
iv. How can you prove that the relationship you suspect between resistance and diameter is correct?
v. The students that produced this data set said, "It took a long time to measure these data, and the Play-Doh was noticeably drier by the end of the lab." Will this influence the validity of the relationship you concluded in part iii? Justify your answer.

264. The figure shows a circuit with a battery of emf ε and negligible internal resistance, and four identical resistors of resistance R numbered 1, 2, 3, and 4. There are three ammeters (A_1, A_2, and A_3) that measure the currents I_1, I_2, and I_3, respectively. The circuit also has a switch that begins in the closed position.

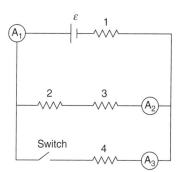

(a) A student makes this claim: "The current I_3 is twice as large as I_2." Do you agree or disagree with the student's statement? Support your answer by writing one or more algebraic equations to support your answer.
(b) Rank the power dissipated into heat by the resistors from highest to lowest, being sure to indicate any that are the same. Justify your ranking.
(c) The power dissipated into heat by resistor 4 is P. Derive an algebraic expression for the power dissipated by resistor 1 in terms of P.

The switch is opened. A student makes this statement: "The power dissipation of resistors 2 and 3 remain the same because they are in parallel with the switch. The power dissipation of resistor 1 decreases because opening the switch cuts off some of the current going through resistor 1."

(d) What parts of the student's statement do you agree or disagree with? Justify your answer with appropriate physics principles and/or mathematical models.

The switch remains open. Resistor 4 is replaced with an uncharged capacitor of capacitance C. The switch is now closed.

(e) i. Determine the current in resistor 1 and the potential difference across the capacitor immediately after the switch is closed.
   ii. Determine the current in resistor 1 and the potential difference across the capacitor a long time after the switch is closed.
   iii. Calculate the energy (U) stored on the capacitor a long time after the switch is closed.

265. A group of students are attempting to determine the internal resistance of a battery. They have connected the battery with an emf of ε and an internal resistance of r with wires to a resistor and meters, as shown in the figure. The voltmeter was positioned to measure the potential difference of the battery, which is called the terminal potential V_T. By changing out the resistor (R), they have collected the data shown in the table.

(Ω)	I (A)	V_T (V)
1	4.8	4.9
2	3.4	6.7
4	2.2	8.6
6	1.6	9.9
10	1.0	10.5
20	0.6	11.1

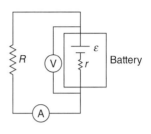

(a) Write an algebraic equation for V_T in terms of ε, r, and I.

From the data, the student graphs the terminal voltage (V_T) as a function of current (I).

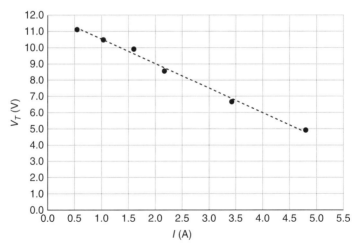

(b) i. Use any of the data available to calculate the emf (ε) and internal resistance (r) of the battery.
ii. Explain the physical meaning of the y-intercept of the best fit line of the graph.
iii. Explain the physical significance of the x-intercept of the best fit line of the graph.

The battery is now used to charge a capacitor with a capacitance of 25 μF by connecting the capacitor to the battery in series.

(c)   i. Will the internal resistance of the battery influence the energy stored in the capacitor? Justify your response.

      ii. A second identical capacitor is connected to the battery in parallel with the first capacitor. Calculate the energy stored in the two-capacitor system.

      iii. The two capacitors connected in parallel are to be replaced by a single new capacitor that will store the same energy. Change a physical variable, the distance between the plates, or the plate area of the original capacitors to accomplish this. Explain your reasoning using principles of physics and/or mathematical models.

CHAPTER 5

Magnetism and Electromagnetic Induction

Skill-Building Questions

266. What are the similarities and differences between electric charges and magnetic poles?

267. Explain what happens when a bar magnet is broken in half.

268. Explain why there are no isolated magnetic poles (monopoles).

269. The north end of a compass always points north. Explain why this happens.

270. Iron filings are sprinkled over the magnet shown in the figure. Sketch the arrangement the filings will take due to the magnetic field.

271. Four compasses are arranged around a magnet, as shown in the figure. Indicate which direction the compasses will turn due to the magnetic field.

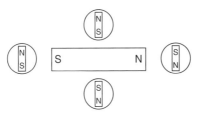

272. A lab table has a long current-carrying wire passing upward through the center. Three compasses are placed around the wire, as shown in the figure. Draw an arrow in each compass indicating in which direction the north end of the compass points.

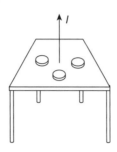

273. Two lightweight, positively charged spheres are suspended from threads near a bar magnet, as shown in the figure. What, if anything, happens to the suspended spheres? Justify your claim.

274. The figure shows a current in a long wire in the plane of the page. Sketch the direction of the magnetic field around the wire.

275. A long wire is carrying a current into the page, as shown in the figure. Indicate the direction of the magnetic field at locations A, B, and C by drawing an arrow. Draw the arrows in proportion to the field strength at that point.

276. Two compasses are positioned so their needles point north in one straight line. A wire is placed over the top of one compass and under the other compass so the wire aligns with the needles, as seen in the figure. How will the compasses be affected when a strong current passes through the wire to the right as shown? Sketch any changes in the direction of the compass heading, and explain your answer.

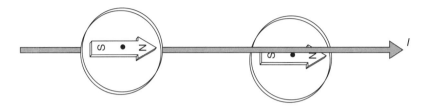

277. Explain the right-hand rule for finding the magnetic field direction around a current-carrying wire.

278. On the axis provided, sketch the magnetic field strength around a current-carrying wire as a function of the perpendicular distance from the wire.

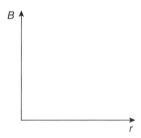

279. Explain the right-hand rule for finding the force on a current-carrying wire in a magnetic field.

280. Draw the direction of the force on the current-carrying wire shown in the magnetic field. The arrow indicates the direction of the current in each case.

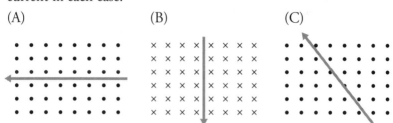

281. Two long wires, a distance (d) apart, carry identical currents in the same direction, as shown in the figure.

(A) What is the direction of the magnetic force on the top wire?
(B) How does the magnetic force on the bottom wire compare in strength and direction to the force on the top wire? Explain.
(C) What is the magnitude and direction of the magnetic field at the midpoint between the wires? Justify your answer.

282. Two long wires, a distance (d) apart, carry different currents in opposite directions, as shown in the figure. The bottom wire has a current twice that of the top wire.

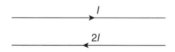

(A) What is the direction of the magnetic force on the top wire?
(B) How does the magnetic force on the bottom wire compare in strength and direction to the force on the top wire? Explain.
(C) Derive an algebraic expression for the magnetic field at the midpoint between the wires.
(D) Derive an algebraic expression for the force per unit length on the top wire by the lower wire.

283. A large horseshoe magnet with a long wire carrying current in the direction of the arrow is shown in the figure.

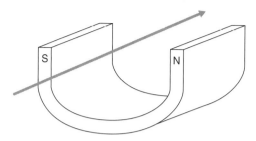

(A) What is the direction of the force on the wire?
(B) What happens to the magnitude of the force on the wire if the current is doubled? Explain your answer.
(C) A rectangular wire connected to a battery is lowered into the large horseshoe magnet, as shown in the figure. What is the direction of the force on the bottom wire in the circuit?

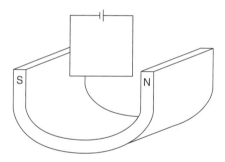

284. Your physics teacher instructs you to investigate the direction of force on moving charges caused by current-carrying wires.
(A) List the items you would use to perform this investigation.
(B) Sketch a simple diagram of your investigation. Make sure to label all items.
(C) Outline the experimental procedure you would use to gather the necessary data. Indicate the measurements to be taken and how the measurement will be used to obtain the data needed. Make sure your outline contains sufficient detail so that another student could follow your procedure and duplicate your results.

285. Explain the differences between gravitational forces due to gravitational fields, electric forces due to electric fields, and magnetic forces due to magnetic fields.

286. Explain the right-hand rule for finding the force on a moving charge in a magnetic field.

287. Two charged particles moving at 4×10^6 m/s through a magnetic field of $B = 0.50$ T are directed out of the page, as shown in the figure.

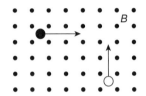

(A) The black circle is a proton moving to the right. Calculate the magnitude and direction of the force on the proton. Show your work.
(B) The white circle is an electron moving toward the top of the page. Calculate the magnitude and direction of the force on the electron. Show your work.

288. Three charged particles move through a magnetic field (B) directed toward the top of the page, as shown in the figure.

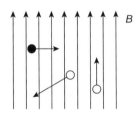

(A) The black circle is a positive charge moving to the right. What is the direction of the force on the charge?
(B) The white circle moving toward the top of the page is a negative charge. What is the direction of the force on the charge?
(C) The white circle moving left and downward is an electron. What is the direction of the force on the electron? Describe the motion of the electron in detail.

289. A particle is moving toward a magnetic field directed into the page, as shown in the figure. Sketch and label the path taken by each of the following particles. Draw all of the pathways in proportion to the paths taken by all other particles. All particles enter the magnetic field with the same initial velocity and direction.

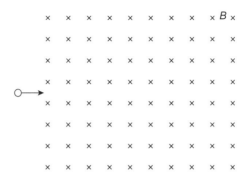

(A) Neutron
(B) Proton
(C) Electron
(D) Positron
(E) Alpha particle

290. Two perpendicular wires carry currents of 1.0 A and 2.0 A, as shown in the figure. Point P is 4 cm from the 1.0 A wire and 5 cm from the 2.0 A wire. Show all your work.

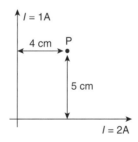

(A) Calculate the net magnitude and direction of the magnetic field at point P.
(B) Calculate the magnitude and direction of the force on a proton placed at point P.
(C) Calculate the magnitude and direction of a force on an electron at point P moving at 3.5×10^6 m/s toward the right.

291. Two long wires carry current perpendicular to the page. One carries 2.0 A of current out of the page, while the other carries 1.0 A into the page, as shown in the figure.

(A) What is the direction of the force on a proton moving to the right through the origin with a velocity of v?
(B) Is there a location along the x-axis where the magnetic field is zero? If so, where is it? If not, explain why not.

292. Four long wires carry identical currents perpendicular to the page, as shown in the figure. The wires are at the corners of a square arrangement. Two wires have current traveling into the page, while the other two carry current out of the page. What is the direction of the magnetic field at point P at the center of the square arrangement?

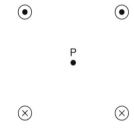

293. A proton traveling toward the top of the page at a velocity of v passes by a magnet, as shown in the figure.

(A) What is the direction of the force on the proton?
(B) What is the direction of the force on the magnet? Explain your answer.
(C) What happens to the force on the proton if the velocity is doubled to $2v$?

294. A particle of mass (m) with a charge of q and traveling at a velocity of v enters a magnetic field B, as shown in the figure. The charge travels in a circle of radius (r) while in the magnetic field.

```
×   ×   ×  B ×

×   ×   ×   ×

×   ×   ×   ×
            ↑v
×   ×   ×   ×
      q ○ m
```

(A) Derive an algebraic expression for the radius of the path the particle takes in the magnetic field as a function of known quantities. Show all your work.
(B) Derive an algebraic expression for the momentum of the particle in terms of r, q, and B. Show all your work.
(C) Derive an algebraic expression for the charge-to-mass ratio in terms of given quantities. Show all your work.
(D) Derive an algebraic expression for the time (T) it takes for the particle to complete one revolution of its circular path in terms of given quantities. Show all your work.

Questions 295–301

A positively charged particle (q) of mass m travels horizontally, with a velocity of v, through the center of two capacitor plates. The plates are separated by a distance of d and connected to a battery of potential difference (ε), as shown in the figure.

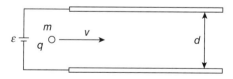

295. Sketch the electric field between the plates.

296. Derive an algebraic expression for the electric field between the plates in terms of given quantities. Show all your work.

297. Describe the motion of the particle as it passes through the capacitor plates. What shape is the path? Which direction is the acceleration?

298. What direction of magnetic field is needed to make the particle travel horizontally straight through the capacitor plates?

299. Derive an algebraic expression for the magnitude of the magnetic field needed to cause this straight, horizontal motion between the plates in terms of given quantities. Show all your work.

300. The crossed electric and magnetic fields are adjusted to cause positively charged particles with a velocity of v to travel straight. What happens to a particle traveling at $2v$? Will it travel straight, or will it curve? If it curves, indicate which way it will curve, and explain why. If it continues to travel straight, explain why.

301. The crossed electric and magnetic fields are tuned to cause positively charged particles with a velocity of v to travel straight. What happens to a *negatively* charged particle traveling at v? Will it travel straight, or will it curve? If it curves, indicate which way it will curve, and explain why. If is continues to travel straight, explain why.

302. In the following figures, a charged particle is passing through a force field. In each case, indicate the direction of the electric or magnetic field that could be used to cause the particle to travel in a straight line.

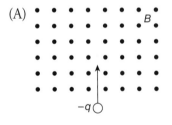

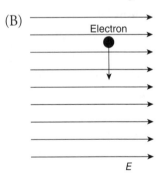

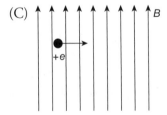

Questions 303–305

A mass spectrometer is a tool used in chemistry to identify the amount and types of chemicals present in a sample by measuring the mass-to-charge ratio of ions. The mass spectrometer has several parts: an ion source, an ion accelerator, a velocity selector, and an ion separator, as shown in the figure.

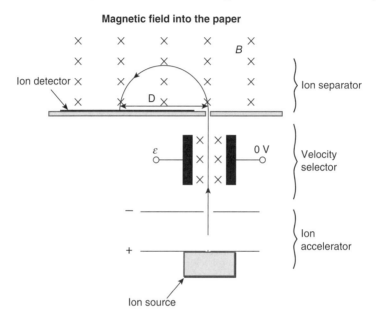

303. An ion with a mass of m and a charge of Q enters the ion accelerator with a velocity of v_1, travels through a potential difference ΔV, and exits at a velocity of v_2. Write an algebraic equation for v_2 as the ion exits the ion accelerator in terms of the given quantities. Show all your work.

304. The ion, traveling at v_2, now enters the velocity selector, which consists of two capacitor plates separated by a distance of d, connected to a battery of potential difference (ε), and with a magnetic field of strength (B). Write an algebraic equation for v_2 of the ion as a function of ε, B, and other known quantities. Show all your work.

305. The ion, still traveling at v_2, enters the ion separator—which has a magnetic field strength of B—and collides with the ion detector at a distance of D from the entry point, as shown in the figure. Write an algebraic equation for D in terms of known quantities. Show all your work.

306. Explain the differences between diamagnetic, paramagnetic, and ferromagnetic materials, making sure to describe the behavior of atomic magnetic fields.

307. What are magnetic domains? Which materials (diamagnetic, paramagnetic, ferromagnetic) exhibit magnetic domains?

308. What happens to magnetic domains in iron when exposed to an external magnetic field?

309. What happens when a strong magnet is held very close above a diamagnetic fluid? What happens when a strong magnet is held very close above a paramagnetic fluid? Explain.

310. What is magnetic flux?

311. Write the equation for magnetic flux, and explain the three ways in which it can be changed.

312. A loop of wire is positioned perpendicular to the end of a magnet, as shown in the figure.

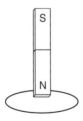

(A) Describe at least two ways to move the wire loop that will produce current in the wire. Explain why each method produces current.
(B) Describe at least two ways to move the magnet so current is produced in the wire. Explain why each method produces current.
(C) Describe at least two methods of how the magnet and/or wire loop can be moved that will not produce current. Explain why neither of these methods produce current.

313. A loop of wire is positioned perpendicular to a magnetic field, as shown in the figure.

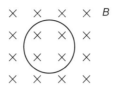

(A) Write the magnetic flux equation. Describe three unique methods for inducing current by changing one of the terms in the magnetic flux equation.
(B) Describe how the wire loop could be moved without inducing an electric current in the wire.
(C) If the wire is held still and the magnetic field increases in strength, which direction will current be induced in the wire?

314. A rectangular loop of wire travels to the right and passes through a region of magnetic field into the page, as shown in the figure.

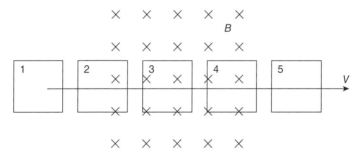

(A) At which locations will current be induced in the wire?
(B) At the locations where current is induced, what direction is the current in the wire loop—clockwise or counterclockwise?

315. Two long parallel wires, separated by a distance of y, pass through a region of magnetic field (B). The two wires are connected by a resistor (R) and a metal bar, separated by a distance of x, to produce a circuit loop, as shown in the figure. The metal bar slides along the wires to the right at a velocity of v.

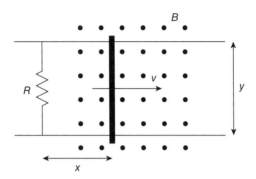

(A) Calculate the induced emf (ε) in the bar in terms of the given quantities.
(B) Calculate the current in the circuit in terms of the given quantities.
(C) What is the direction of the current in the resistor—upward or downward?

316. Explain how a dynamic microphone and an electric generator use electromagnetic induction.

AP-Style Multiple-Choice Questions

317. A proton is moving toward the top of the page when it encounters a magnetic field that changes its direction of motion. After encountering the magnetic field, the proton's velocity vector is pointing out of the page. What is the direction of the magnetic field? Assume gravitational force is negligible.

    (A) Toward the bottom of the page
    (B) To the right
    (C) To the left
    (D) Into the page

318. An electron is moving downward toward the bottom of the page when it passes through a region of magnetic field, as shown in the figure by the shaded area. The electron travels along a path that takes it through the spot marked X. The gravitational force on the electron is very small. What is the direction of the magnetic field?

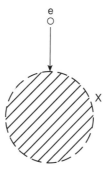

(A) Toward the bottom of the page
(B) Toward the top of the page
(C) Out of the page
(D) Into the page

Questions 319 and 320

A proton is moving at a velocity of 6.0×10^4 m/s to the right, in the plane of the page, when it encounters a region of magnetic field with a magnitude 0.12 T perpendicular to the page, as shown in the figure.

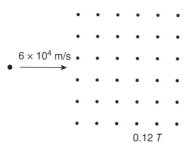

319. Which of the following is the radius of curvature of the path of the proton?
 (A) 5×10^1 m
 (B) 5×10^{-3} m
 (C) 5×10^{-5} m
 (D) 5×10^{-7} m

320. The proton is replaced with an electron moving in the same direction and at the same speed. Which of the following best describes the deflection direction and the radius of curvature of the electron in the magnetic field?

Deflection direction	Radius of curvature
(A) Same as proton	Larger than proton's
(B) Same as proton	Smaller than proton's
(C) Opposite of proton	Larger than proton's
(D) Opposite of proton	Smaller than proton's

321. In each of the answer choices below, either a proton or an electron is moving toward the top of the page through either an electric or a magnetic field. In which case does the charged particle experience a force to the right? *Select two answers.*

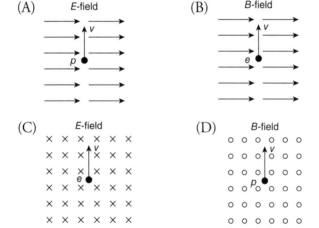

322. Two long parallel wires carry currents (I_A and I_B), as shown in the figure. Current I_A in the left wire is twice that of current I_B in the right wire. The magnetic force on the right wire is F. What is the magnetic force on the left wire in terms of F?

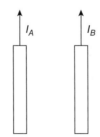

(A) F in the same direction
(B) F in the opposite direction
(C) $F/2$ in the same direction
(D) $F/2$ in the opposite direction

323. An iron magnet is broken in half at the midpoint between its north and south ends. What is the result?

    (A) A separate north pole and south pole, each with the same magnetic strength as the original magnet
    (B) A separate north pole and south pole, each with half the magnetic strength of the original magnet
    (C) Two separate north-south magnets, each with the same magnetic strength as the original magnet
    (D) Two separate north-south magnets, each with half the magnetic strength of the original magnet

324. The figure shows the microscopic dipoles inside two metal objects. Copper is diamagnetic. Iron is ferromagnetic. Which of the following best depicts the microscopic internal dipole position when the objects are placed in a strong, external magnetic field directed toward the top of the page?

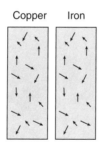

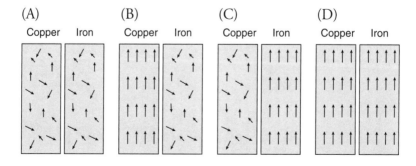

325. Compasses are arranged in a tight circle around a long wire that is perpendicular to the plane of the compasses. The wire is represented in the figures by a dot. The wire carries a large current directly into the page. Which of the following best depicts the orientation of the compass needles?

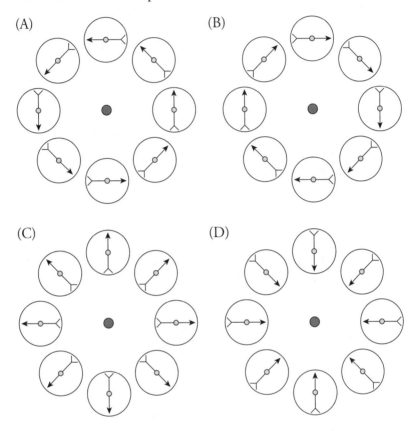

326. A magnetic field, directed into the page, is placed between two charged capacitor plates, as shown in the figure. The magnetic and electric fields are adjusted so a proton moving at a velocity of v will pass straight through the fields. The speed of the proton is doubled to $2v$. Which of the following force diagrams most accurately depicts the forces acting on the proton when traveling at $2v$?

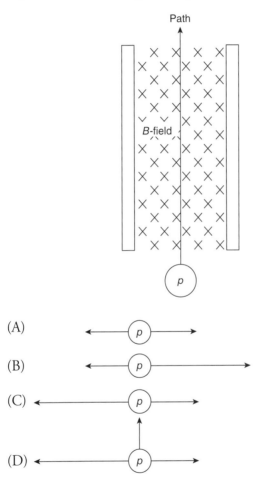

Questions 327 and 328

An electron is traveling at a constant speed of v parallel to a wire carrying a current of I, as shown in the figure. The electron is a distance of d from the wire.

327. Which of the following is true concerning the force on the current-carrying wire due to the electron?

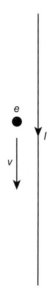

(A) The force is directed toward the right.
(B) The force is directed toward the left.
(C) The force is directed into the page.
(D) There is no force on the current-carrying wire due to the electron.

328. The force on the electron from the current is F. Which of the following will increase the force to $2F$? *Select two answers.*
(A) Halve the distance of the electron to the wire.
(B) Halve the velocity of the electron.
(C) Double the current in the wire.
(D) Double the current in the wire and halve the distance of the electron to the wire.

329. A dynamic microphone contains a magnet and a coil of wire connected to a movable diaphragm, as shown in the figure. Sound waves directed at the diaphragm generate a current in the wires leading from the coil. Which of the following helps to explain why this occurs?

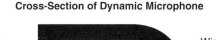

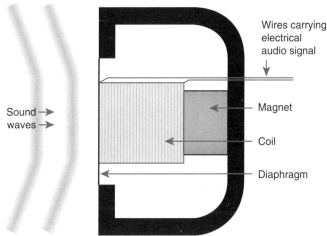

(A) The area of the coil changes.
(B) The magnitude of the magnetic field produced by the magnet changes.
(C) The angle between the plane of the coil and the magnetic field produced by the magnet change.
(D) The strength of the magnetic field in the plane of the coil changes.

AP-Style Free-Response Questions

330. The apparatus shown in the figure consists of a container on the left filled with air that has been heated until it produces ions of gas molecules and released electrons. Some of the particles exit the hole in the right side of the container and accelerate through a potential difference of ε provided by a capacitor. The capacitor has a hole in the center of the right plate through which the particles enter a region of magnetic field with a field strength of B, represented by the shaded area of the figure. Assume that the gravity force is negligible compared to other forces.

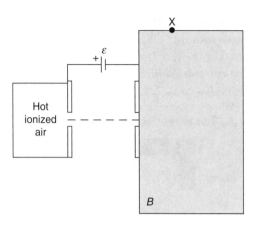

(a) Briefly explain why positive ions reach the region of magnetic field, but that electrons do not.

(b) A singly ionized nitrogen molecule, with a charge of $+e$ and a mass of m_N, exits the gas container and accelerates across the capacitor plates, reaching a velocity of v when it enters the region of magnetic field.
  i. Derive an algebraic expression of the velocity of the nitrogen ion in terms of given quantities. Assume the initial velocity of the ion is small.
  ii. After entering the region of magnetic field, the nitrogen ion passes through the point marked X. On the figure, sketch the path of the nitrogen ion.
  iii. Derive an algebraic expression for the radius of the nitrogen ion path in terms of given quantities.

(c) The potential difference of the battery is doubled. Demonstrate how this will affect the path of the nitrogen ion by placing the letter Y at the boundary of the magnetic field where the nitrogen ion will pass through.

(d) The potential difference of the battery is returned to its original value. A doubly ionized nitrogen molecule with a charge of $+2e$ enters the capacitor from the heated air container. Based on your equations from part (b), describe the motion of the doubly ionized nitrogen through the capacitor and magnetic field in comparison to the motion of the singly ionized nitrogen molecule.

331. A lead box containing radioactive materials that emit both electrons and positrons is placed near an apparatus consisting of an evacuated capacitor that is filled with a magnetic field, as shown in the figure. Electrons that enter along the center line of the capacitor plates travel straight through (undeflected) with a velocity of $v = 1.0 \times 10^7$ m/s and out the hole in the center of the apparatus on the right. The capacitor plates are separated by a distance of $d = 0.020$ m; each plate has an area of $A = 1.0 \times 10^{-4}$ m² and a potential difference of ΔV. A uniform magnetic field of $B = 0.030$ T is directed out of the page between the plates, as shown in the figure.

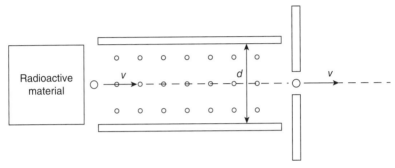

(a) Explain why it is acceptable to neglect the effects of gravity on the electrons passing through the apparatus.

(b) i. Explain why the electrons pass through the capacitor plates undeflected. Support your argument with an algebraic equation and an appropriately drawn force diagram.
  ii. Use your equation to calculate the potential difference (ΔV) between the capacitor plates.
  iii. Which capacitor plate has the highest potential? Justify your reasoning making reference to the electric field.
  iv. Calculate the magnitude of the energy that is stored in the capacitor.

(c) A positron enters the apparatus along the same path as the electrons from part (b).
  i. Explain why the positron, traveling at the same speed as the electrons, will also travel straight through the device undeflected. Support your argument with an equation.
  ii. A second positron enters the apparatus at a speed of $2v$. Sketch the path of the positron through the capacitor plates on the figure.

(d) An electron exits the apparatus at a velocity of $v = 1.0 \times 10^7$ m/s parallel to a long wire of a circuit, as shown in the figure. The distance between the electron and the wire is 1 mm.

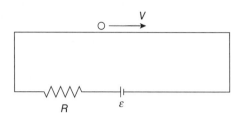

i. Calculate the potential difference-to-resistance ratio $\left(\frac{\varepsilon}{R}\right)$ of the circuit such that the electron will experience a force F of 1.3×10^{-16} N.
ii. Draw a force vector on the figure to show the direction of the force on the electron.

CHAPTER 6

Geometric and Physical Optics

Skill-Building Questions

332. Describe what a wave is.

333. Explain the difference and similarities between longitudinal and transverse wave vibration.

334. Sketch a representation of a longitudinal, a transverse, and an electromagnetic (EM) wave, including the direction of the oscillation and direction of propagation.

335. Explain the differences and similarities between EM waves and mechanical waves.

336. Describe what each term in the following equations represents:

$$E = A\cos(2\pi ft) = A\cos\left(\frac{2\pi t}{T}\right)$$

$$B = A\cos\left(\frac{2\pi x}{\lambda}\right)$$

337. For each of the following electromagnetic wave representations, write the appropriate wave equation.

(A)

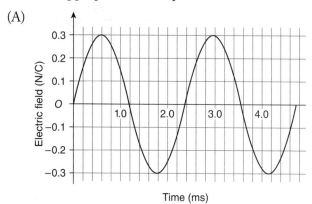

(B)

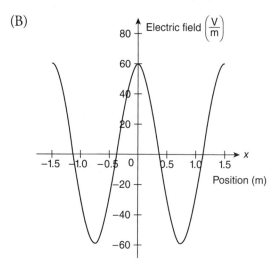

(C)

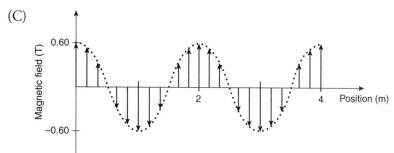

338. Sketch each of these wave equations on the axis:

(A) $B = 4.0\cos\left(\dfrac{\pi x}{2}\right)$, where B is in units of T, and x is in units of m.

(B) $E = 2\sin(1.05t)$, where E is in units of N/C, and t is in units of s.

339. Explain wave superposition.

340. Explain constructive and destructive interference.

341. Two waves travel toward each other, as shown in the figure. Sketch at least three unique interference patterns that will be seen as the waves pass each other.

342. Explain how the Doppler effect is displayed in EM waves. Give an example of how this physics phenomena is put to use in technology.

343. If all waves are vibrations that transport energy, what does an EM wave "vibrate"? Explain why this allows an EM wave to travel through the vacuum of space.

344. How do we know that EM waves are transverse?

345. Use the "picket fence and rope" analogy, as shown in the figure, to explain EM wave polarization.

346. In the figure, nonpolarized light is approaching a polarizing filter. The filter only lets through light that is aligned with the filter axis.
(A) Explain why a single polarizer will let through only 50 percent of the nonpolarized light that tries to pass through.

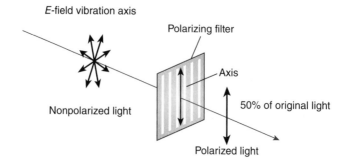

(B) How much light will get through this arrangement of polarizing filters?

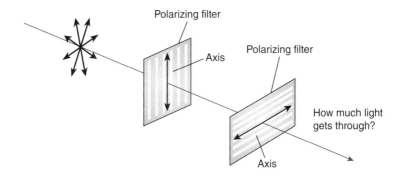

(C) Why does light get through this arrangement of polarizing filters?

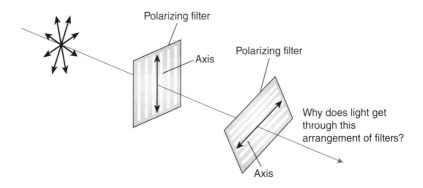

347. List the EM spectrum from longest wavelength to shortest.

348. What are the similarities and differences between radio waves and X-rays?

349. How fast does visible light travel in a vacuum? How fast do microwaves travel in a vacuum?

350. What is the wavelength range of visible light?

351. What is the frequency range of visible light?

352. Explain the connection between wave fronts and light rays. Draw a sketch of wave fronts and light rays that would be produced by a point light source and by a flat light source like a TV screen.

353. Explain Huygens' principle and point source model, and use the model to sketch how a plane wave and a circular wave front propagate forward.

354. Use the point source model of wave propagation to describe and explain the behaviors of the waves listed here. Sketch a representation to assist in your explanation.

(A) Diffraction of a plane wave front as it passes by a boundary (see figure).

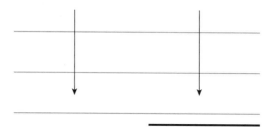

(B) Sound waves can be heard around corners, but light waves do not seem to bend around corners. Why is this?

(C) Diffraction of a plane wave front as it passes through a small opening comparable in size to the wavelength (see figure).

(D) Diffraction of a plane wave front as it passes through an opening that is larger than the wavelength (see figure).

(E) Double-slit interference pattern that appears when a plane wave front passes through two small openings (see figure).

(F) Refraction of light as light passes from air straight into a block of glass (see figure). Which way will the light turn and why?

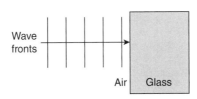

(G) Refraction of light as light passes at an angle from a faster speed medium into a slower speed medium (see figure). Which way will the light turn and why?

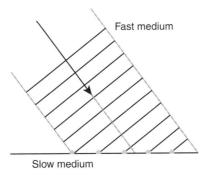

(H) A plane wave front of light reflecting off a flat mirror (see figure).

355. Green light waves are projected toward a barrier with a narrow slit of width W, which is comparable in width to the wavelength of light being used. This produces an alternating light-dark pattern on a screen, as shown in the figure. The barrier is a distance of Y from the screen. The light source is a distance of X from the barrier. For each of the following modifications to the setup of the apparatus, describe the changes that will be observed in the light pattern seen on the screen, and sketch the new pattern. Justify each claim with an equation.

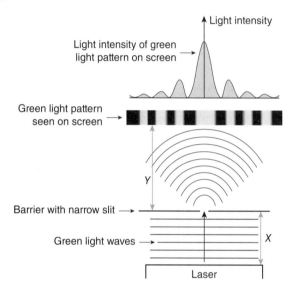

(A) Increase X.
(B) Increase Y.
(C) Increase W.
(D) Use a red laser instead of a green one.
(E) Use a violet laser instead of a green one.

356. Use an equation to explain why sound waves traveling through a doorway will bend around the corners of the door and fill the room beyond with sound, but light waves will not bend around the corner and fill the room with light.

357. Green light waves are projected toward a barrier with two narrow slits of width W separated by a distance of Z. This produces an alternating light-dark pattern on a screen, as shown in the figure.

The barrier is a distance of Y from the screen. The light source is a distance of X from the barrier. For each of the following modifications to the apparatus, describe the changes that will be observed in the light pattern seen on the screen, and sketch the new pattern. Justify each claim with an equation.

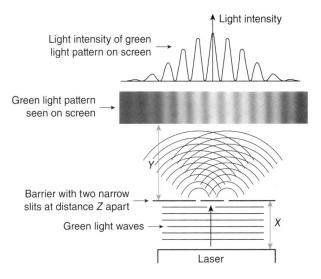

(A) Decrease W.
(B) Decrease X.
(C) Decrease Y.
(D) Decrease Z.
(E) Use a red laser instead of a green one.
(F) Use a violet laser instead of a green one.
(G) Replace the double-slit barrier with a multislit diffraction grating with the same slit spacing of Z.

358. The figure shows monochromatic light passing through two narrow slits separated by a distance of d and striking a screen at a distance of l away on the right. A pattern of light and dark spots appears on the screen. In a clear, coherent, paragraph-length response, explain why there will be a bright spot on the screen in the two examples on the left and dark spots on the screen in the example on the right. Be sure to reference the wave property of interference in your explanation, and explain the significance of the term ΔL.

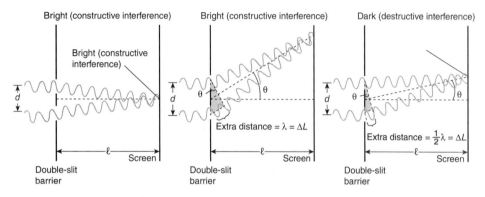

359. How is the interference pattern created by a double slit similar and different from that produced by a multislit diffraction grating?

360. In a clear, coherent, paragraph-length response, explain the phenomenon of thin film interference patterns. Include in your response a discussion of how the thickness of the thin film required to produce constructive interference is affected by phase changes when the wave reflects off the surface.

361. State the law of reflection, and draw a figure representing it.

362. A light ray travels to the right toward two mirrors that are at right angles to each other, as shown in the figure below. The normal line at the first angle of incidence is shown. Sketch the path of the light ray as it reflects off the two mirrors. Be sure to draw the normal line at the point of incidence on the second mirror, and mark all angles of incidence and reflection.

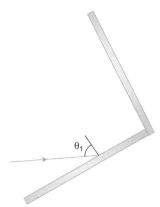

363. A cat is looking at a mouse in a mirror. Draw two rays showing where the image of the mouse would appear to the cat. Include normal lines and angles of incidence and reflection.

364. Describe the three phenomena that can occur to light when it encounters a boundary between transparent materials. Explain what impact the similarities or differences in the materials have on these behaviors. Include an explanation of how energy is conserved.

365. Discuss how the velocity, frequency, and wavelength change when light passes from one transparent material into another, with a greater index of refraction.

366. Rank the speeds of light in these substances: glass, vacuum, water, and air.

367. Yellow light ($\lambda = 580$ nm) traveling through air encounters the calm surface of a swimming pool ($n = 1.33$).
    (A) Calculate the speed of light in the pool water.
    (B) Calculate the wavelength of the light in the pool water.
    (C) Calculate the frequency of the light in the pool water.
    (D) Sketch a ray diagram of the yellow light as it encounters the pool. Be sure to show both the reflected and refracted portions of the light. Mark the normal and all angles.

Questions 368–371

Students shine a laser beam at a rectangular block of acrylic plastic, as shown in the figure. Two dots of light appear on a screen above the acrylic block, one to the right and one to the left. Likewise, two dots of light appear below the block, one to the right and one to the left.

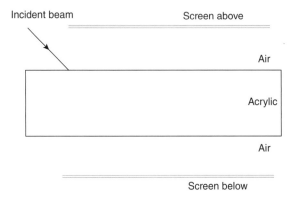

368. Sketch a ray diagram that could produce such an arrangement of light dots. In your diagram indicate which angles measured between the light ray and the normal line are equivalent.

369. Which of the dots (right or left) is brighter than the other on the screen both above and below the acrylic block? Explain your reasoning.

370. The original acrylic block is replaced by a new block with a rectangular air gap in the middle, as shown in the figure. The laser beam positioned at #1 does not produce a dot on the top or bottom screen. Explain why this is, and draw a ray diagram to support your claim.

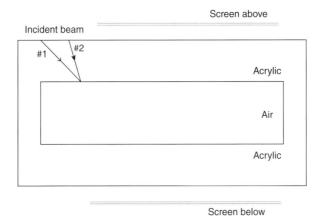

371. When positioned at #2, the laser produces a dot of light on both screens. Explain why this happens, and draw a ray diagram to support your claim.

372. A beam of light strikes the interface between water ($n = 1.33$) and air at an angle of incidence of 45°, as shown in the figure.

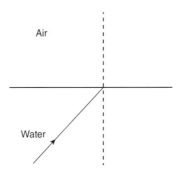

(A) Sketch the ray diagram indicating the angles of reflection and refraction.
(B) Calculate the angles of reflection and refraction. Show your work.
(C) At what range of angles would the beam of light not enter the air? Show your work.
(D) Is it possible to send a beam of light from the air toward the surface of the water in such a way that none of the light enters the water, but instead, 100 percent reflects off the surface of the water? Justify your claim with a diagram and an equation.

373. Your physics teacher instructs you to determine the index of refraction of a triangular glass prism.
   (A) List the items you would use to perform this investigation.
   (B) Sketch a simple diagram of your investigation. Make sure to label all items, and label the measurements you would make.
   (C) Outline the experimental procedure you would use to gather the necessary data. Indicate the measurements to be taken and how the measurements will be used to obtain the data needed. Make sure your outline contains sufficient detail so that another student could follow your procedure and duplicate your results.

374. A group of students are given a semicircular sapphire prism through which they shine a beam of light, as shown in the figure. They measure the incident and refracted angle of the light and produce the table shown.

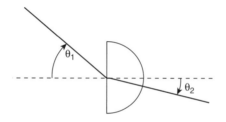

Incidence angle (degree)	Refraction angle (degree)	$\sin \theta_i$	$\sin \theta_r$
10	6	0.17	0.10
20	11	0.34	0.19
30	16	0.50	0.28
40	21	0.64	0.36
50	26	0.77	0.43
60	29	0.87	0.49
70	32	0.94	0.53
80	34	0.98	0.56

   (A) Use the data to determine the index of refraction of the prism.
   (B) Graph the refraction angle as a function of incidence angle. Are the two variables directly related? Justify your claim.

(C) Graph the sine of the refraction angle as a function of the sine of the incidence angle. Use the slope of the graph to calculate the index of refraction of the prism. Show your work.

375. An incident ray of light strikes the surface of a triangular prism, as shown in the figure. Light exits all three sides of the prism.

(A) Explain how this is possible.
(B) Sketch the path of the light ray as it travels through the prism to show how light can exit the prism at three locations. Mark all normal lines, making sure the angles you draw are proportionally correct.
(C) Complete your drawing by sketching any reflected rays that you have not already drawn on the figure. Make sure the angles you draw are proportionally correct.

376. An incident ray of light strikes the surface of a triangular prism at 30°, as shown in the figure.

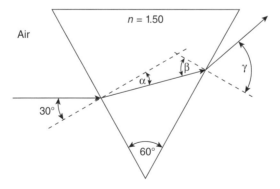

(A) Calculate angle α.
(B) The light ray strikes the right side of the prism at angle β. Does the light ray exit the prism as shown in the figure? Support your claim with an equation and a mathematical argument.

377. A glass prism has a point at the bottom made up of 45° angled surfaces, as shown in the figure. A light beam directed downward through the top of the prism is completely reflected off the bottom surfaces and exits back through the top of the prism. When submerged in water, the light beam exits the bottom of the prism.

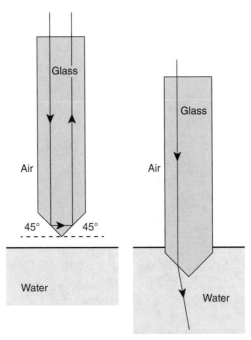

(A) Calculate the minimum index of refraction of the glass.
(B) In a clear, coherent, paragraph-length response, explain why light exits the bottom of the prism when it is submerged in water. Your explanation should discuss the speed of light.

378. Light strikes a glass sphere, as shown in the figure. Point C is the center of the sphere. Complete the ray diagram showing all reflected and transmitted rays, the normal lines, and indicating the measure of all angles.

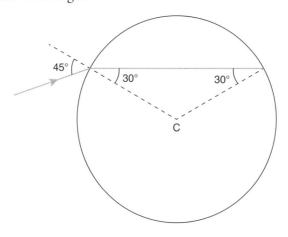

379. Explain how fiber optics work.

380. Explain the differences between real and virtual images. How can an image be tested to determine whether it is real or virtual?

381. What is the mathematical relationship between the radius and the focal length of a spherical mirror?

382. What is the significance of the focal point for mirrors and lenses?

383. What kind of lens or mirror is each of those shown in the figure: concave/convex, converging/diverging? Explain.

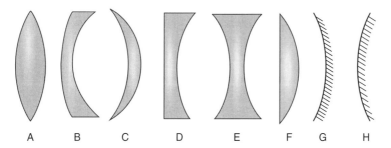

384. Explain how to draw the three principal rays for both concave/convex lenses on optical ray diagrams.

188 › 500 AP Physics 2 Questions to know by test day

Questions 385–389

The figures show two spherical mirrors, each with a radius of R, and two thin lenses with a focal length f. Complete the trajectories of each ray according to the descriptions in each question.

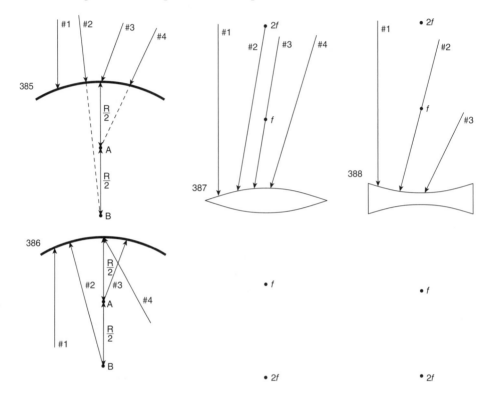

385. Four light rays are directed at a convex mirror. Points A and B are spaced half a radius from the mirror as shown.

386. Four light rays are directed at a concave mirror. Points A and B are spaced half a radius from the mirror as shown.

387. Four rays are directed at a thin convex lens with a focal length of f.

388. Three rays are directed at a thin concave lens with a focal length of f.

389. The lenses and mirrors are submerged in water. Would this affect the paths of the rays you just drew? Explain why or why not.

390. A mirror and a lens, both with a focal length of f, are submerged in oil. Do the focal lengths of either change? If so, how does the focal length change? Justify your answer.

391. The figure shows three light rays approaching two separate optical devices.

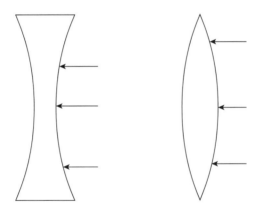

(A) Complete the paths of the rays, and use the rays to locate the focal points. Be sure to draw the normal lines wherever the rays pass through a lens surface.
(B) The thickness of each lens is increased without changing its diameter. What, if anything, happens to the focal length of the lenses?

392. An object is placed in front of each of the spherical mirrors shown in the figure.

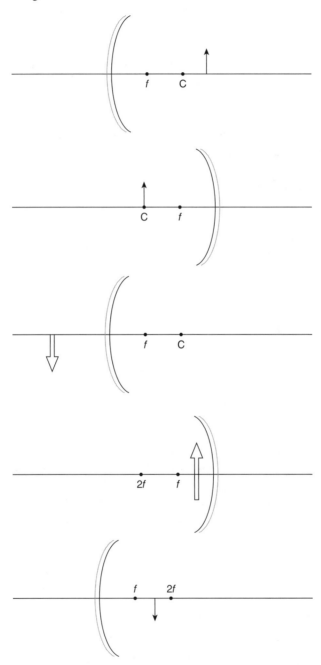

(A) Sketch a ray diagram with at least two rays to locate the position of the image. Sketch the image in each case.
(B) Next to each image, indicate whether the image is real or virtual, inverted or upright, and enlarged or reduced in size.
(C) Sketch a stick figure on each diagram to indicate where a person would have to be standing and in which direction the person needs to look to see the image formed by the mirror. The small figure here shows an example of a person looking to the left.

(D) What distinguishes a real image from a virtual image on the ray diagrams? Explain.

393. An object is placed in front of each of the thin lenses shown in the figure.

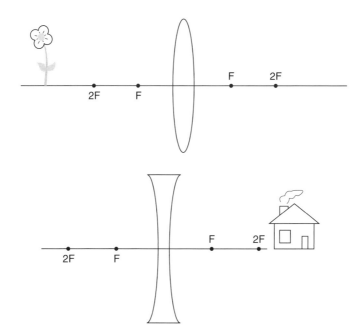

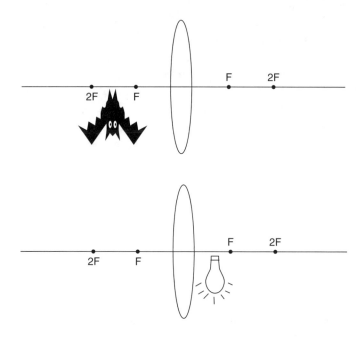

(A) Sketch a ray diagram with at least two rays to locate the position of the image. Sketch the image.
(B) Next to each image, indicate whether the image is real or virtual, inverted or upright, and enlarged or reduced in size.
(C) Sketch a stick figure on each diagram to indicate where a person would have to be standing and in which direction the person needs to look to see the image formed by the mirror.
(D) What distinguishes a real image from a virtual image in these ray diagrams? Explain.
(E) What would happen to the image of the flower if the bottom half of the lens were covered by a piece of cardboard? Justify your claim by making reference to the ray diagram.

394. Complete the missing data for the optical device in the table below.

Type of optical device	Object location	Focal length (+ or −)	Image location (+ or −)	Real or virtual	Inverted or upright	Image magnification
Diverging lens	Between F and 2F	−	− Same side of lens			
Convex lens			+ Between F and 2F		Inverted	Smaller
Converging lens			+ Beyond 2F	Real		
Convex lens		+		Virtual		Larger
Converging mirror	Beyond 2F					Smaller
Concave mirror	Between F and 2F					
Converging mirror			− Behind the mirror		Upright	
Convex mirror	Between F and the mirror					Smaller

395. A concave mirror is taken outside on a sunny day. The mirror is held so sunlight falls directly on the mirror's surface. This procedure produces a bright dot of light 20 cm from the surface of the mirror on a screen. Calculate the focal length and radius of the mirror.

396. A lens with a focal length of 10 cm produces an inverted image 30 cm from the lens. What is the magnification of the image, and where was the object located?

397. A dentist wants a mirror that will produce an upright image that is 2.5 times larger when placed 5 cm from a tooth. What kind of mirror will do this, and what is its radius of curvature?

398. A diverging lens has a focal length of 40 cm. If an object is 20 cm from the lens, what is the image distance and magnification of the image?

399. In the human eye, the distance from the lens to the retina, on which images are focused, is about 1.7 cm. If a book is held 30 cm from the eye, what should the focal length of the eye be?

400. A diverging mirror with a radius of curvature of 50 cm produces an image that is 15 cm from the mirror. What is the object location, and what is the magnification of the image?

401. Your physics teacher instructs you to determine the focal length of a concave mirror.
  (A) List the items you would use to perform this investigation.
  (B) Sketch a simple diagram of your investigation. Make sure to label all items, and indicate measurements you would need to make.
  (C) Outline the experimental procedure you would use to gather the necessary data. Indicate the measurements to be taken and how the measurements will be used to obtain the data needed. Make sure your outline contains sufficient detail so that another student could follow your procedure and duplicate your results.
  (D) Could this procedure be used to find the focal length of a convex mirror? Justify your response.

402. Your physics teacher instructs you to determine the focal length of a convex lens.
  (A) List the items you would use to perform this investigation.
  (B) Sketch a simple diagram of your investigation. Make sure to label all items, and indicate measurements you would need to make.
  (C) Outline the experimental procedure you would use to gather the necessary data. Indicate the measurements to be taken and how the measurements will be used to obtain the data needed. Make sure your outline contains sufficient detail so that another student could follow your procedure and duplicate your results.
  (D) Could this procedure be used to find the focal length of a concave lens? Justify your response.

Questions 403 and 404

A lens lab produces these data and the graph.

s_o	s_i	$1/s_o$	$1/s_i$
50	12.5	0.020	0.080
30	15.0	0.033	0.067
25	16.7	0.040	0.060
20	20.0	0.050	0.050
15	30.0	0.067	0.033
12	60.0	0.083	0.017

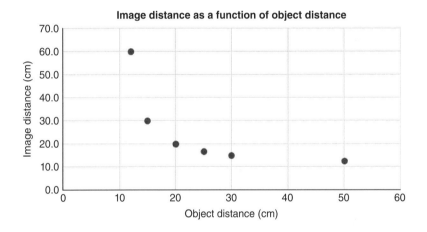

403. What data would you plot to produce a straight line? Plot the data.

404. What information from your straight line plot will allow you to determine the focal length of the lens? Justify your claim with an equation. Calculate the focal length of the lens using your straight line graph.

405. Students take a crystal ball to the beach. They notice that when looking through the crystal ball, the image of the beach, sun, and sky are inverted, as shown in the figure.

(A) The crystal ball is behaving like what kind of lens? Back up your claim with evidence.
(B) Draw a ray diagram that shows how the image could be inverted by the crystal ball.

406. The human eye produces an inverted image on the retina, as shown in the figure.

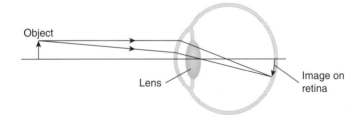

(A) What type of lens does the human eye have? Justify your claim.
(B) Some people have an eye disorder called myopia (nearsightedness) in which the image forms in front of the retina. What corrective lens is needed to fix this problem? Justify your answer.
(C) Some people have an eye disorder called hyperopia (farsightedness) in which the image forms behind the retina. What corrective lens is needed to fix this problem? Justify your answer.

AP-Style Multiple-Choice Questions

407. A lens and a mirror both have a focal length of f in air. Both are submerged in water and the focal length f_{water} is measured for both. How does the focal length under water compare to the focal length in air?

	Lens	Mirror
(A)	$f = f_{water}$	$f = f_{water}$
(B)	$f = f_{water}$	$f < f_{water}$
(C)	$f < f_{water}$	$f = f_{water}$
(D)	$f < f_{water}$	$f < f_{water}$

408. Which of the following correctly describes the motion of the electric and magnetic fields of a microwave transmitted by a cell phone?
    (A) Both the electric and magnetic fields oscillate in the same plane and perpendicular to the direction of wave propagation.
    (B) Both the electric and magnetic fields oscillate perpendicular to each other and to the direction of wave propagation.
    (C) The electric field oscillates perpendicular to the direction of wave propagation. The magnetic field oscillates parallel to the direction of wave propagation.
    (D) Both the electric and magnetic fields oscillate parallel to the direction of wave propagation.

409. In a laboratory experiment, you shine a green laser past a strand of hair. This produces a light and dark pattern on a screen. You notice that the lab group next to you has produced a similar pattern on a screen, but the light and dark areas are spread farther apart. Which of the following could cause the light and dark pattern to spread? *Select two answers.*
    (A) The second group used thinner hair.
    (B) The second group is using a red laser.
    (C) The second group had the screen closer to the hair.
    (D) The second group held the laser farther from the hair.

410. An observer can hear sound from around a corner but cannot see light from around the same corner. Which of the following helps to explain this phenomenon?
    (A) Sound is a longitudinal wave, and light is an electromagnetic wave.
    (B) Sound is a mechanical wave, and light is a transverse wave.
    (C) Light travels at a speed much faster than that of sound.
    (D) Light has a much smaller wavelength than sound.

411. Which of the following best represents the electric field (E) measured in mV/m as a function of time measured in nanoseconds (ns)?

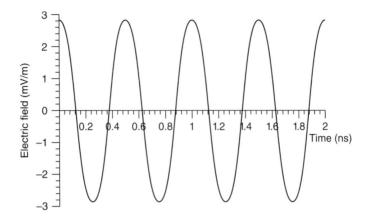

(A) $E = 6\cos(25.1t)$
(B) $E = 6\cos(12.6t)$
(C) $E = 3\cos(12.6t)$
(D) $E = 3\cos(25.1t)$

412. A mirror produces an upright image one-half the height of the object when the object is 12 cm from the mirror's surface. What is the focal length of the mirror?
(A) −12 cm
(B) −4 cm
(C) 4 cm
(D) 6 cm

413. A light ray with a wavelength of λ_w and a frequency of f_w in water ($n = 1.33$) is incident on glass ($n = 1.61$). In the glass, the wavelength and frequency of the light are λ_g and f_g. How do the values of wavelength and frequency of the ray of light in water compare to those in glass?

	Wavelength	Frequency
(A)	$\lambda_w > \lambda_g$	$f_w = f_g$
(B)	$\lambda_w > \lambda_g$	$f_w > f_g$
(C)	$\lambda_w < \lambda_g$	$f_w = f_g$
(D)	$\lambda_w < \lambda_g$	$f_w < f_g$

414. An optics bench is set up on a meter stick, as shown in the figure. The light source is a candle placed at x_0. The lens is located at x_1. The screen is moved until a sharp image appears at location x_2. The lens and screen are moved multiple times to produce crisp images of differing sizes. The data for each x_0, x_1, and x_2 location is recorded in a table. Which of the following procedures will allow a student to determine the focal length of the lens?

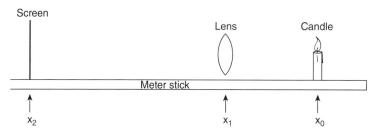

(A) Plot x_2 as a function of x_0. The focal length will be the vertical axis intercept.
(B) Plot $(x_2 - x_1)$ as a function of $(x_0 - x_1)$. The focal length will be the vertical axis intercept.
(C) Plot $1/x_2$ as a function of $1/x_0$. The focal length will be the inverse of the vertical axis intercept.
(D) Plot $1/(x_2 - x_1)$ as a function of $1/(x_0 - x_1)$. The focal length will be the inverse of the vertical axis intercept.

415. A laser beam passes through a prism and produces a bright dot of light a distance of x from the prism, as shown in the figure. Which of the following correctly explains the change in distance x as the angle (θ) of the prism is decreased?

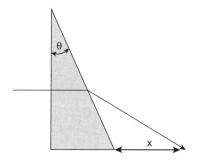

(A) The distance x increases because the angle on incidence increases.
(B) The distance x increases because the angle of incidence decreases.
(C) The distance x decreases because the angle on incidence increases.
(D) The distance x decreases because the angle of incidence decreases.

416. Which of the following could be the path of a light ray passing through a glass prism with an index of refraction of *n* = 1.5? *Select two answers.*

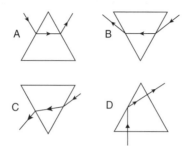

417. In the human eye, the distance from the lens to the retina, on which the image is focused, is 20 mm. A book is held 30 cm from the eye, and the focal length of the eye is 16 mm. How far from the retina does the image form, and what lens should be used to place the image directly on the retina?

	Distance of image from retina	Corrective lens
(A)	3.1 mm in front of the retina	Concave lens
(B)	3.1 mm in front of the retina	Convex lens
(C)	14 mm behind the retina	Concave lens
(D)	14 mm behind the retina	Convex lens

AP-Style Free-Response Questions

418. In a laboratory experiment, an optics bench consisting of a meter stick, a candle, a lens, and a screen is used, as shown in the figure. A converging lens is placed at the 50-cm mark of the meter stick. The candle is placed at various locations to the left of the lens. The screen is adjusted on the right side of the lens to produce a crisp image. The candle and screen locations on the meter stick produced from this lab are given in the table. Extra columns are provided for calculations if needed.

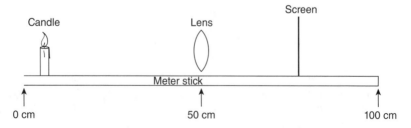

Candle location (cm)	Screen location (cm)				
0	71				
10	74				
20	80				
25	88				
29	100				

(a) Calculate the focal length of the lens using one of the data points in the table. Show your work.

(b) Use the data to produce a straight line graph that can be used to determine the focal length of the lens. Explain how you found the focal length from the graph.

(c) i. Sketch a ray diagram to show how the candle would produce an upright image with a magnification larger than 1.0. Draw the object, at least two light rays, and the image. Indicate the locations of the focus (f and $2f$).

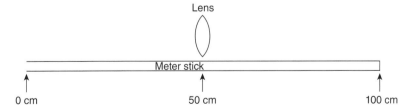

ii. A student says that virtual images can be projected on a screen. Do you agree with this claim? How could you perform a demonstration to support your stance with evidence?

419. The figures show two different representations of the same plane wave traveling through medium #1 and approaching a boundary between two transparent media. The left figure shows a light ray representation. The right figure shows a wave front representation. The index of refraction of medium #1 is greater than that of medium #2.

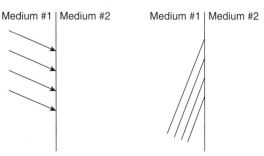

(a) i. On the left figure, complete the diagram by sketching the path of all four rays in medium #2.
  ii. On the right figure, complete the diagram by sketching all four wave fronts in medium #2.
(b) The figure represents another series of plane waves traveling toward and incident on a barrier in its path. One wave model treats every point on a wave front as a point source.

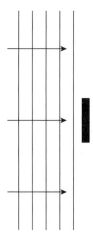

  i. Use the point source model to help you sketch the wave as it passes the barrier.
  ii. In a clear, coherent paragraph-length response, describe how this point source explains the shape of the wave as it passes the barrier and why an interference pattern is produced beyond the barrier on the right.

(c) The figure represents another series of plane waves traveling toward and incident on a barrier with two identical openings.

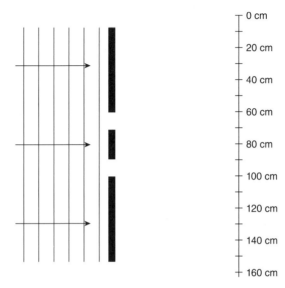

i. Sketch a representation of the interference pattern produced on the wall beyond the barrier. The wall is marked with centimeters for your reference. The 80-cm mark is aligned with the center of the barrier. Indicate locations of constructive interference with the letter C and locations of destructive interference with the letter D.
ii. The distance between the wave fronts is increased. How does this affect the interference pattern?

420. A student is given a semicircular glass prism and a laser. The student directs the laser perpendicular to the curved surface, as shown in the figure.

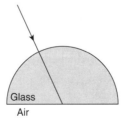

(a) i. Sketch the path of the ray exiting the prism.
ii. Explain the path of the light exiting the prism, making reference to the speed of light in the glass and air.

(b) The index of refraction of the glass can be found by graphing a straight line. Indicate what quantities should be graphed to produce a straight line graph and how the graph could be used to determine the index of refraction of the glass.

(c) Outline an experimental procedure that could gather the data necessary to determine the index of refraction of the glass. Include sufficient detail so that another student could follow the procedure. Besides the semicircular glass prism and the laser, list any additional equipment to be used. Indicate any angles to be measured on the figure above.

(d) In another experiment, the laser is directed at a rectangular block of glass, as shown in the figure. When the laser is directed along path A, the light exits both the top and left side of the glass block. When directed along path B, the light only exits the left side of the glass block.

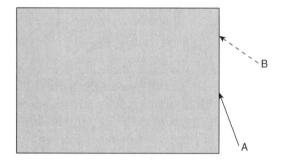

  i. Explain why the light behaves differently depending on the path.
 ii. Using a solid line, sketch the light along path A. Be sure to show enough detail so it is apparent how the light exits both the top and the left side of the block.
iii. Using a dashed line, sketch the light along path B. Be sure to show enough detail so it is apparent how the light exits the left side of the block.
iv. Which path will produce the brightest beam exiting the left side of the block? Justify your reasoning.

CHAPTER 7

Quantum, Atomic, and Nuclear Physics

Skill-Building Questions

421. A car traveling at half the speed of light has its headlights and taillights on.
    (A) What speed does the driver measure for the speed of the headlights and taillights?
    (B) What speed does a person standing on the side of the road measure for the speed of the headlights and taillights?

422. Explain the two postulates of special relativity.

423. Under what conditions do we need to consider the implications of special relativity?

424. Give an example when two observers will not agree on time and length.

425. Explain what the photoelectric effect is.

426. Explain why ultraviolet (UV) light can discharge a negatively charged electroscope but not a positively charged electroscope.

427. Sketch a picture of the photoelectric experiment. Explain how the energy of ejected electrons is determined.

428. Explain the following terms from the experiment: *photon energy*, *stopping potential*, *threshold/cutoff frequency*, and *work function*.

429. Explain the characteristic results of the photoelectric experiment that support the photon (particle) model of light. Discuss why the wave model of light cannot produce the results of the experiment.

430. What factors determine the energy of a photon of light? Write the equation for photon energy.

431. List the electromagnetic (EM) spectrum from highest to lowest energy. Also, list the visible light spectrum from highest to lowest energy.

432. What is an eV (electron volt)?

433. Sodium has a work function of 2.4 eV. Iron has a work function of 4.7 eV. Cesium has a threshold frequency of 4.7×10^{14} Hz.
    (A) Calculate the threshold frequency of sodium and iron. Calculate the work function of cesium.
    (B) Plot photoelectron energy as a function of photon frequency for the three elements. Label the slope.
    (C) Ultraviolet light (200 nm) shines on cesium. What is the maximum energy of ejected electrons?

434. Which of the following behaviors of light support the wave model of light, and which support the particle model of light? Justify your response in each case.
    (A) Light that passes through a double slit produces an alternating pattern of intensity.
    (B) X-rays can be used to ionize gas.
    (C) X-rays can be directed at crystals to produce interference patterns.
    (D) Photons of light can impart momentum to electrons in a collision.
    (E) Long wavelength radio waves can bend around hills and buildings.
    (F) Light from distant stars is redshifted.
    (G) Infrared (IR) light does not generate electricity in a solar cell no matter how intense the light is.

435. Are photons affected by electric and magnetic fields the way electrons and protons are?

436. If photons of light behave like particles, do they have mass and momentum like other particles? Explain.

437. What are the equations for photon energy and momentum? Show how the two equations are connected by the speed of light.

438. A photon moving to the right collides with a stationary electron, as shown in the figure. The photon recoils to the left along the same path.

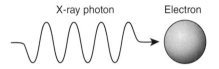

(A) Describe any changes to the photon. Justify your claims.
(B) Describe any changes to the electron. Justify your claims.

439. A photon moving to the right collides with a stationary electron. The electron moves off at an angle, as shown in the figure.

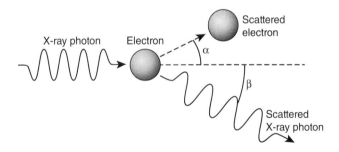

(A) Describe any changes to the photon. Justify your claims.
(B) Is this collision elastic? Explain your reasoning.
(C) Write the equations of conservation of momentum and conservation of energy for this collision.

440. Sketch the Rutherford model of the atom, and explain its structure.

441. Explain why the planetary model (electrons orbiting the atom like planets orbiting around the sun) cannot be correct.

442. The Bohr model of the atom has electrons in "stable states." Discuss the significance of stable states in this model of the atom.

443. Explain how electrons transfer between stable states in the Bohr model of the atom. Discuss both upward and downward transitions.

Questions 444–450

The figure shows an energy-level diagram of an atom.

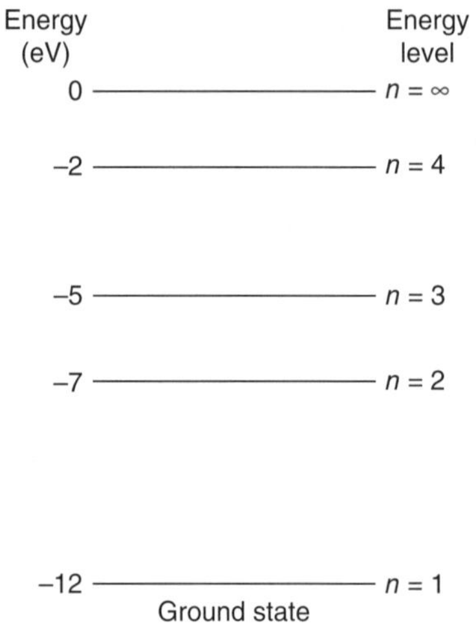

Energy (eV)	Energy level
0	$n = \infty$
−2	$n = 4$
−5	$n = 3$
−7	$n = 2$
−12	$n = 1$ Ground state

444. What happens when a 7-eV photon strikes an electron in the ground state? Draw the transition in the diagram, if possible, and discuss any changes to the energy of the atom.

445. What happens when a 9-eV photon strikes an electron in the ground state? Draw the transition in the diagram, if possible, and discuss any changes to the energy of the atom.

446. What happens when a 15-eV photon strikes an electron in the ground state? Draw the transition in the diagram, if possible, and calculate the kinetic energy of the electron.

447. What are all the possible photon emission energies from an electron starting in the $n = 4$ energy level (the third excited state)?

448. An electron transitions between the following energy levels:

$$n = 4 \text{ to } n = 3$$
$$n = 4 \text{ to } n = 2$$
$$n = 3 \text{ to } n = 2$$
$$n = 2 \text{ to } n = 1$$

   (A) Rank the emitted photons from highest to lowest frequency.
   (B) Rank the emitted photons from longest to shortest wavelength.
   (C) Does the atom gain or lose energy during these transitions? Justify your answer.

449. An electron transitions between the following energy levels:

$$n = 1 \text{ to } n = 3$$
$$n = 2 \text{ to } n = 4$$
$$n = 3 \text{ to } n = \infty$$

   (A) Rank the absorbed photons from highest to lowest frequency.
   (B) Rank the absorbed photons from longest to shortest wavelength.
   (C) Does the atom gain or lose energy during these transitions? Justify your answer.

450. Calculate the energy, wavelength, and frequency of the photon that is emitted by an electron dropping from the $n = 4$ to the $n = 3$ energy level. Is this photon visible? Explain.

451. Visible light shines on an unknown gas. The gas is found to absorb light of a 400-nm wavelength. When the light is turned off, the gas is seen to emit both 400-nm and 600-nm wavelengths of light.
   (A) Draw an energy level diagram for this gas. Label the energy levels in electron volts.
   (B) Is there another wavelength emission that is not in the visible spectrum? Justify your answer with your diagram.

Questions 452–455

The figure shows a material used in a laser. The lasing material consists of atoms with a metastable state at energy level E_2. Photon #1 is used to pump electrons to energy level E_3, where they quickly fall to level E_2. This is shown in the laser pumping phase of the figure. When stimulated by another photon of identical energy, the electrons that have been sitting in the metastable E_2 energy level cascade down toward E_1, producing a laser beam, as shown in the lasing phase of the figure.

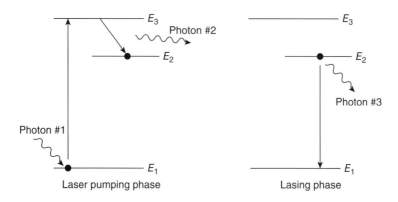

452. What happens to the overall energy of the atom during the laser pumping and lasing phases? Justify your answer by using the figure.

453. Write an equation for the energy gain or loss of the atom in the laser pumping phase.

454. Write an equation for the energy of photon #3 in terms of the energy levels of the atom.

455. Write an equation that relates the energies of the three photons (#1, #2, and #3).

456. What is a de Broglie wave?

457. Sketch the de Broglie wavelength of an electron as a function of momentum and velocity on the two provided axes.

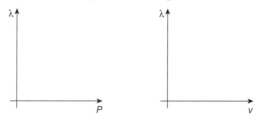

458. What is the evidence that particles, such as electrons and protons, exhibit wave properties?

459. Electron beams produce diffraction patterns when directed at crystals, but a stream of baseballs directed at a crystal do not. Why?

460. How do the wave properties of electrons help to explain Bohr's stable electron states? Be sure to reference the idea of wavelength and standing waves in your explanation.

461. A gun fires electrons at a barrier with two small openings. A screen registers electron impact locations as shown in the figure.

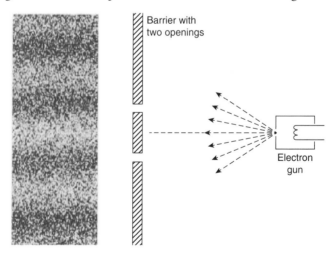

(A) What does this prove about the nature of electrons?
(B) What happens to the pattern when the voltage used to accelerate the electrons is increased? Justify your answer with an equation.
(C) What happens to the pattern if the distance between the openings is decreased? Justify your answer with an equation.

462. What is a wave function? What does it represent?

463. The two figures represent the wave functions of different particles.

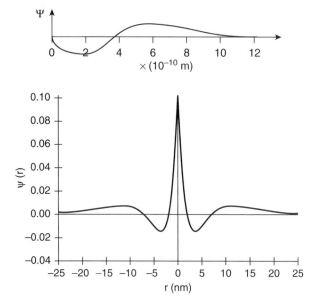

(A) On each of the figures, circle the locations where the particles are most likely to be found. Rank the locations from most likely to least likely.
(B) Place an X on the horizontal axis of each figure where the particle will never be found.

464. Scientists in the early 1900s discovered that some materials (radioactive substances) emitted strange, unknown rays, originally called Becquerel rays. In an effort to understand the nature of these rays, scientists sent them through electric fields and observed the results. The figure shows the path of three rays that exit a box of radioactive material.

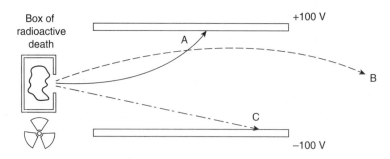

(A) Determine the direction of the electric field through which the rays are being sent. Justify your claim.
(B) What can be learned about rays A, B, and C from this experiment? Explain.
(C) What could each of the rays be? Justify your answer.

465. A lead box of radioactive material has an opening, so that the escaping radiation is directed to the right, as shown in the figure. The radiation passes through a region of crossed E- and B-fields so that only a specific speed of charged particles can travel straight through. Sketch the path of each of the following. Make sure that each path is proportional to the others.

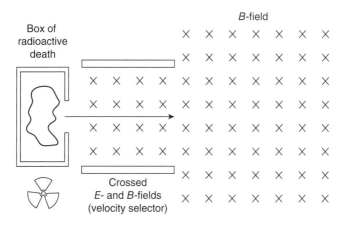

(A) Alpha particle
(B) Beta particle
(C) Gamma ray
(D) Proton
(E) Neutron
(F) Electron
(G) Positron
(H) Hydrogen nucleus
(I) Helium nucleus

466. What is the atomic mass unit? When do we use this unit?

467. Complete the following table of properties for subatomic particles:

Particle	Symbol	Mass (u)	Mass (kg)	Charge (e)	Charge (C)
Proton		1.0073 u	1.673×10^{-27}	+e	1.6×10^{-19}
Electron		0.0005 u	9.11×10^{-31}	–e	-1.6×10^{-19}
Neutron		1.0087 u	1.675×10^{-27}		
Alpha particle		4.0012 u	6.644×10^{-27}		
Beta particle					
Gamma ray					
Positron					
Hydrogen nucleus					
Helium nucleus					

468. What are the parts of an atom? Sketch a simple picture of how they are arranged. Your sketch should show approximate sizes.

469. Discuss the forces that affect electrons and cause them to be part of the atomic structure.

470. Discuss the forces that the nucleus experiences. Explain how the nucleus stays together.

471. What is an isotope? Give examples.

472. Not all isotopes are stable. Explain why not. Give examples of stable and unstable isotopes.

473. Do different isotopes of the same atom have the same chemical and nuclear properties? Explain.

474. What is mass defect, and how is it related to both the nuclear strong force and binding energy?

475. Discuss the similarities and differences between the nuclear binding energy and the ionization energy of an electron.

476. Write a mathematical equation for the mass defect (Δm) of a helium nucleus.

477. An iron nucleus ($^{56}_{26}$Fe) has a mass of 92.86×10^{-27} kg (55.94 u). Find the mass defect and binding energy of this nucleus. Show your work.

478. List the four conservations that must be obeyed during nuclear reactions. Explain each in detail.

479. Explain the mechanics of alpha, beta, and gamma decay, referencing the four conservations for each.

480. Why does nuclear decay occur?

481. How does the mass of the reactants compare to the mass of the products during decay? Explain.

482. Explain the mechanics of nuclear fission and fusion with reference to the four conservations.

483. Decide whether the following nuclear reactions are possible. If not, explain why not.
    (A) $^{6}_{3}\text{Li} + ^{4}_{2}\text{He} \rightarrow ^{12}_{7}\text{N} + 2^{0}_{-1}\beta + \text{energy}$
    (B) $^{3}_{1}\text{H} + ^{2}_{1}\text{H} \rightarrow ^{4}_{2}\text{He} + ^{1}_{1}\text{H} + \gamma$

484. Complete the following reactions, and classify each as either decay (specify the type), fission, or fusion.
    (A) $2^{1}_{1}\text{H} + 2^{1}_{0}\text{n} \rightarrow X + \text{energy}$
    (B) $^{137}_{55}\text{Cs} \rightarrow X + ^{0}_{-1}e$
    (C) $^{238}_{92}\text{U} \rightarrow ^{234}_{90}\text{Th} + X + \text{energy}$
    (D) $^{1}_{0}\text{n} + ^{235}_{92}\text{U} \rightarrow ^{144}_{54}\text{Xe} + ^{90}_{38}\text{Sr} + X$
    (E) $^{12}_{6}\text{C} \rightarrow ^{12}_{6}\text{C} + X$

485. Use the figures to estimate the half-life of the two radioactive samples.

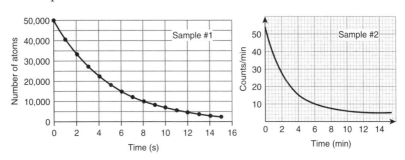

486. Explain what the term *half-life* means. Be sure to include reactants and products in your discussion.

487. The half-life of carbon-14 is 5,730 years. Fresh wood has about 12.6 carbon-14 decays/min/g. A piece of wood from an archaeological dig has a radioactivity of 1.6 decays/min/g. How old is the wood?

488. A room is filled with 20 kg of the radioactive gas radon-222. Radon decays by remitting an alpha particle and transmuting into a polonium, as shown in this equation: $^{222}_{86}Rn \rightarrow {}^{218}_{84}Po + \alpha +$ energy. Radon-222 has a mass of 222.018 u and a half-life of 3.8 days. Polonium-218 has a mass of 218.009 u and a half-life of 3.10 minutes. Alpha particles have a mass of 4.002 u.

(A) Calculate the energy released in the decay of one radium atom. Show your work.

(B) Assume that all the energy released during one decay is converted into the kinetic energy of the alpha particle. What will be the recoil velocity of the polonium atom? Show all your work.

(C) How many kilograms of radon gas remain after 19 days?

(D) How much polonium is there after 19 days? Explain. (Careful, this is a trick question!)

AP-Style Multiple-Choice Questions

489. An astronaut in a rocket is passing by a space station at a velocity of 0.33 c. Looking out the window, the astronaut sees a scientist on the space station fire a laser at a target. The laser is pointed in the same direction that the astronaut is traveling. On which of the following observations will the astronaut and scientist agree?

(A) The length of the rocket.
(B) The time it takes the laser to hit the target.
(C) The speed of the laser beam.
(D) The astronaut and scientist will not agree on any of these measurements.

490. In an experiment, neon gas atoms are ionized by high-energy photons. The ionized gas is then accelerated to a uniform velocity and directed into a magnetic chamber, as shown in the figure. Approximately 90 percent of the ionized atoms follow path #1, which has a radius of 0.029 m, while about 9 percent of the neon ions follow path #2, which has a radius of 0.032 m. Which of the following is the most reasonable explanation for this phenomenon?

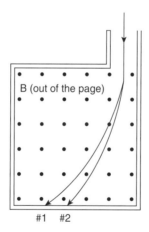

(A) The ions are exhibiting wave properties of constructive and destructive interference.
(B) The ions have different numbers of electrons.
(C) The ions have different numbers of protons.
(D) The ions have different numbers of neutrons.

491. The graph shows the wave function of a particle as a function of x in the region between -25 nm $< x < +25$ nm. At which of the following positions is the probability of finding the particle greatest?

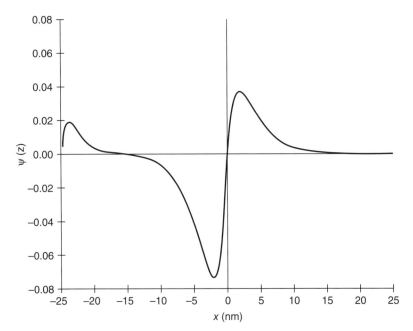

(A) -24 nm
(B) -15 nm and 0.0 nm
(C) -2 nm
(D) 2 nm

492. A nucleus of $^{237}_{93}$Np goes through a sequence of decays during which it emits four beta particles and some alpha particles to finally end up as a stable $^{205}_{81}$Tl nucleus. How many alpha particles have been emitted in this process?

(A) 32
(B) 26
(C) 8
(D) 4

493. Under the right conditions, a photon of light can be converted into a positron and electron, as shown in the figure and represented in the following equation: $\gamma \rightarrow e^+ + e^-$. Which of the following correctly explain why this interaction is possible? *Select two answers.*

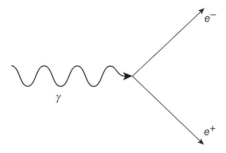

(A) The positron is the antimatter particle of the electron, so the mass of the two particles cancel.
(B) The signs of the positron and the electron cancel.
(C) The momentum of the two particles sum to that of the photon.
(D) The kinetic energy of the two particles sum to that of the photon.

494. The figure shows energy levels for a gas. Electrons in the ground state are excited by a continuous range of electrical energy from 100 eV to 115 eV. After the electrical input is turned off, the gas emits light in discrete frequencies. Which of the following correctly indicates the electron transition from the gas that would produce the highest frequency light?

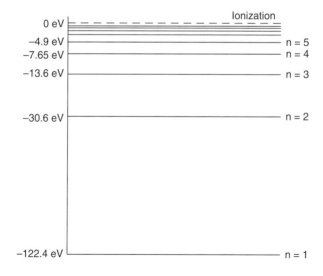

(A) $n = 5$ to $n = 4$
(B) $n = 4$ to $n = 3$
(C) $n = 3$ to $n = 2$
(D) $n = 5$ to $n = 2$

495. A uniform ultraviolet light source shines on two metal plates, causing electrons to be emitted from both plates. The two plates are made of different materials but have the same surface area. Plate A emits more electrons than plate B. However, the electrons emitted from plate B have a higher kinetic energy. Which of the following describe plausible explanations for the differences in electron emissions? *Select two answers.*
    (A) Plate A has a larger work function than plate B.
    (B) Higher energy electrons from plate B would be produced by placing the plate closer to the light source, where it would receive more ultraviolet photons from the source.
    (C) More electrons would be produced from plate A by placing the plate closer to the light source, where it would receive more ultraviolet photons from the source.
    (D) Plate A emits more electrons of lesser energy, while plate B emits fewer electrons of higher energy, but the total combined energy of the emitted electrons is the same.

496. Which of the following phenomena involving electrons can be explained by the wave model of matter? *Select two answers.*
    (A) The photoelectric effect
    (B) Atomic energy levels
    (C) Crystal diffraction
    (D) Beta decay

497. Which of the following explains why the nucleus of a stable atom is bound together?
    (A) The gravitational force between the neutrons and protons is greater than the repulsive electric force between the protons.
    (B) The neutrons polarize and create an attractive electric force that cancels the repulsive electrostatic force of the protons.
    (C) The orbit of electrons creates a magnetic force on the protons that is greater than the repulsive electric force.
    (D) The strong force between nucleons is greater than the repulsive electric force of the protons.

$$^{12}_{6}C + ^{1}_{1}H \rightarrow ^{13}_{7}N + \gamma$$

498. Which of the following correctly expresses the energy of the released gamma ray in the reaction represented by the equation shown?
   (A) $m_C + m_H - m_N$
   (B) $(m_C + m_H - m_N)c^2$
   (C) $(m_C + m_H - m_N)c^2 - hf_\gamma$
   (D) $(m_C - m_H - m_N)c^2$

AP-Style Free-Response Questions

499. A radioactive lithium nucleus is at rest when it decays as shown in the following reaction:

$$\text{Li} \rightarrow \text{He} + \text{p} + 2.475 \text{ MeV energy}$$

   Helium has a mass of 4.0026 u. Protons have a mass of 1.0073 u.
   (a) How many protons and neutrons does the original lithium have? Explain how you arrived at these numbers, making reference to conservation laws.
   (b) The helium nucleus is stable. Explain how this is possible by making reference to the forces involved.

   The 2,475 MeV of energy released in this reaction is in the form of kinetic energy of the helium and proton.
   (c) In a few sentences, explain where this kinetic energy comes from. Write an algebraic equation to support your explanation.
   (d) In a few sentences, discuss the movement of the helium and proton after the reaction. Write an algebraic equation to support your explanation.
   (e) Calculate the rest mass of the lithium nucleus.

   The half-life of this isotope of lithium is 370×10^{-24} s.
   (f) Scientists want to use this lithium isotope in a research investigation. Could researchers expect to find this isotope on Earth, or would they need to manufacture it in the laboratory? Explain your reasoning.

500. A gas with a ground state of –8.0 eV is illuminated by a broad spectrum of ultraviolet light and is found to absorb the 248 nm wavelength of light. When the ultraviolet light is turned off, the gas sample emits three different wavelengths of light: 248 nm, 400 nm, and 650 nm.

(a) On the axis provided, construct and label an energy level diagram that displays the process of both the absorption and emissions by the gas. Show your supporting calculations below.

eV

The light emitted by the gas is directed toward a sample of tin. It is found that the 248-nm emission from the gas causes electrons to be ejected from the tin, but that the 400-nm emission does not.

(b) Will the 650-nm emission eject electrons from the tin? If so, explain how it could be accomplished. If not, explain why it is not possible.

(c) The light from the gas is directed at a sample of potassium that subsequently ejects electrons with a maximum energy of 2.71 eV.
   i. Calculate the de Broglie wavelength of the maximum energy electrons.
   ii. These electrons are directed at two small openings spaced 2 nm apart. Will this result in the formation of an interference pattern? Justify your answer using appropriate physics principles.

ANSWERS

Chapter 1: Fluids

Skill-Building Questions

1. The density stays the same at ρ. The volume and mass will go up eight times: $8V$ and $8m$.

2. (A) Use a spring scale to measure the weight of the block. Convert the weight to mass: $F_g = mg$.

Use a ruler to measure the length, width, and height. Calculate the volume of the block. Calculate density: $\rho = \dfrac{m}{V}$.

(B) You can get the mass of the rock as before, but the volume is now an issue. That's because a ruler cannot accurately measure the rock's dimensions to obtain the volume. You could find the volume by water displacement. Fill a beaker to the brim. Submerge the rock and catch the overflow water. The volume of displaced water should equal the volume of the rock.

3. (A) See the graph.

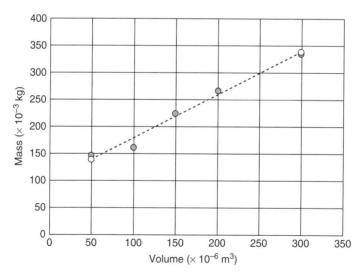

(B) You forgot to tare/zero the balance. The intercept is the mass of the graduated cylinder (approximately 100 g).

(C) The density of the fluid equals the slope of the best fit line of the data. The slope is about 800 kg/m³.

‹ 223

(D) More data points over a wider range almost always helps. Notice that the volume is measured to the nearest 50 ml. So, you could also use a more accurate graduated cylinder that measures more precisely. Taring or zeroing the balance won't really help because the slope of the line is independent of the mass of the graduated cylinder.

4. The wood has a density that is one-third that of water. The buoyancy force upward must equal the weight of the wood. Remembering that mass is equal to density times volume ($m = \rho \times V$), we can derive the following relationship:

$$F_g = F_B$$
$$(mg)_{wood} = (\rho V g)_{water}$$
$$(\rho V g)_{wood} = (\rho V g)_{water}$$
$$\rho_{wood} V_{wood} = \rho_{water} V_{water}$$
$$\rho_{wood} V_{wood} = \rho_{water}\left(\frac{1}{3}V_{wood}\right)$$
$$\rho_{wood} = \frac{1}{3}\rho_{water}$$

5. Atoms in a fluid are in constant motion. They collide with the walls of the container, bounce off, and impart a tiny force on the wall due to their change in momentum during the collision. The sum of all these atomic collisions adds up to a net force over an area (pressure) on the walls of the container.

6. (A) $P_B = P_C > P_A = P_D$. The pressure depends on the height of the fluid: $P = P_0 + \rho g h$.
   (B) $F_B = F_C > F_A = F_D$. Force equals pressure times area: $F = PA = (P_0 + \rho g h)A$.
   (C) $P_C > P_A = P_B = P_D$. The pressure depends on the height of the fluid above the stopper: $P = P_0 + \rho g h$.

7. Pascal's principle states that any increase in pressure on the surface of a fluid creates an equal and undiminished increase in pressure on all points throughout the fluid.

Questions 8–10

8. We don't have to worry about pressure changes due to height differences because the pump and cylinder 3 have the same height. Pascal's principle tells us that the pressure from the pump will be transmitted throughout the fluid. Therefore,

$$P_{pump} = P_3$$

$$\frac{F_{pump}}{A_{pump}} = \frac{F_3}{A_3} = \frac{F_3}{4A_{pump}}$$

$$F_3 = 4F_{pump}$$

9. Both cylinders receive the same pressure increase due to the pump. However, the top of cylinder 3 is higher than the top of cylinder 1, so the pressure will be greater for cylinder 1.

10. The pressure at the top of cylinders 1 and 2 will be the same because they are the same height and have the same amount of fluid above them in the pump.

Questions 11–13

11. The water pressure at the hole is determined by the height of the water above the hole. As the amount of water decreases, the pressure that pushes the water out of the hole decreases as well. In terms of energy, think of the amount of fluid above the tack as gravitational potential energy that can be converted to kinetic energy in the water exiting the tack hole. As the water level descends, the gravitational potential energy decreases, which decreases the kinetic energy of the exiting water.

12. When the water level descends, the air pocket at the top of the bottle increases in volume. With the cap on the bottle, air from the outside cannot flow in. The ideal gas law, $PV = nRT$, tells us that as the volume of air above the water increases, the pressure of that air must decrease. The static pressure equation, $P = P_0 + \rho g h$, tells us that as the pressure on top of the water decreases, the water pressure at the tack hole will also decrease. Eventually, the water pressure inside the bottle equals the air pressure outside the bottle, and we have equilibrium. The water stops coming out of the tack hole.

13. When water stops flowing out of the bottom tack hole, the pressure inside the water and the atmospheric pressure are the same—equilibrium. The top tack is at a location with less fluid pressure because it is higher in the fluid. When the top tack is removed, the atmospheric pressure of the air is higher than the liquid pressure inside the bottle at that spot. Air will push into the hole and bubble to the top of the bottle, increasing the pressure above the water; this pushes water out of the bottom tack hole. Simply put, air enters the top tack hole, and water exits the bottom tack hole.

Questions 14 and 15

14. As the piston is pulled upward, the volume of the air gap increases. This decreases the pressure in the air gap: $PV = nRT$. This causes a decrease in the water pressure at the bottom of the cylinder: $P = P_0 + \rho g h$. The atmospheric pressure on top of the lake is still the same. Therefore, there is a greater pressure in the lake water than in the water at the bottom of the cylinder. Therefore, water is pushed up the cylinder until the pressure at the bottom of the cylinder is in equilibrium with the water pressure of the lake. Note that water is NOT sucked up the tube.

15. There is a limit to how high a fluid can be drawn up in this fashion ($P = P_0 + \rho g h$). The lowest pressure we can have at the top of the water column is a vacuum, $P_0 = 0$. Therefore, the water can be drawn to its highest point when there is a complete vacuum at the top of the cylinder.

16. A barometer works by having a long sealed tube filled with fluid, usually mercury, and a vacuum at the top. Air pressure keeps the fluid in the tube. The height of the fluid is dependent on the density of the fluid and the atmospheric pressure:

$$P_{atmosphere} = \rho g h$$

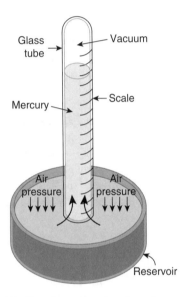

Thus, the height of the fluid is directly proportional to the atmospheric pressure.

Questions 17 and 18

17. The gas pressure is less than that of the atmosphere because the atmosphere is able to push the oil down farther than the trapped gas can.

18. The surface oil pressure on the right side is equal to atmospheric pressure. A point horizontal to this one is also equal to atmospheric pressure because static fluid pressure is constant along a horizontal:

$$P_{atm} = P_{gas} + \rho g h$$

$$P_{gas} = P_{atm} - \rho g h$$

$$P_{gas} = 100{,}000 \text{ Pa} - (930 \text{ kg/m}^3)(9.8 \text{ m/s}^2)(0.22 \text{ m}) = 98{,}000 \text{ Pa}$$

19. Static fluid pressure increases with depth. An object in a fluid will, therefore, have a greater static pressure at the bottom than at the top. This difference in pressure over the surface area of the object creates the buoyancy force. This also directs the buoyancy force upward against gravity.

20. $F_B = (\rho V)_{\text{displaced fluid}}(g) = (m)_{\text{displaced fluid}}(g) =$ weight of displaced fluid

21. (A) When a person floats in a pool, he or she receives a buoyancy force upward from the water. By Newton's third law, the water must also receive an equal and opposite force downward from the person. This downward force increases the force on the bottom of the pool.

    (B) When a person floats in a pool, he or she displaces water, which makes the water level rise. This rise in water height causes an increase in the static fluid pressure on the bottom of the pool, which in turn increases the force on the bottom of the pool.

22. To float, an object must displace water equal to the object's weight. If the object is denser than water, this is not possible, and the object sinks.

$$(mg)_{\text{object}} = (\rho V g)_{\text{object}} > (\rho V g)_{\text{water}}$$

23. A ship is not made completely of steel. It has air pockets inside that bring its average density below that of water. This is why ships sink when they develop a hull breach. The air pockets are filled with water, the average density of the ship rises above that of water, and the ship sinks.

24. (A) The spring scale reading decreases because the buoyancy force pushes up on the mass. The balance reading increases because the mass pushes down on the water with a Newton's third law force that is equal in magnitude but opposite in direction to the upward buoyancy force on the mass.

    (B) Both scales' readings remain the same. The object does not displace any additional water moving from position 2 to position 3. Therefore, the buoyancy force does not change by moving the mass from position 2 to position 3.

25. Your friend is correct! When the cargo is inside the raft, the raft must support the weight of the cargo. When the cargo is thrown overboard, the cargo will displace water and supply some additional buoyancy force. This means that the raft does not have to support all of the weight of the cargo and can hold more people without sinking.

26. (A) $P = P_{\text{atm}} + \rho g z_2$
    (B) $F_B = \rho V g = \rho(xyz_2)g$

(C) The portion of the raft above water will be used to support the additional weight:

$$F_B = \rho V g = \rho(xyz_1)g = Mg$$
$$M = \rho(xyz_1)$$

Questions 27–29

27. The vector lengths of gravity plus tension must add up to the length of buoyancy, as shown in the figure.

28. $\Sigma F = 0$
$F_{Buoy} = mg + T$
$T = F_{Buoy} - mg$
$T = \rho V g - mg$ But: $m_{wood} = \rho V_{wood}$
$T = \rho V g_{Fluid} - \rho V g_{wood}$ Since the block is completely submerged $V g_{Fluid} = V g_{wood}$
$T = Vg(\rho_{Fluid} - \rho_{wood})$
$T = (0.1 \times 0.1 \times 0.2 \text{ m}^3)(9.8 \text{ m/s}^2)(1000 \text{ kg/m}^3 - 600 \text{ kg/m}^3)$
$T = 7.8 \ N$

29. With the string cut $\Sigma F = ma$

$F_{Buoy} - mg = ma$

$\rho V g - mg = ma$ But $m_{wood} = \rho V_{wood}$

$\rho V g_{Fluid} - \rho V g_{wood} = \rho V a_{wood}$ Note that V cancels

$(1000 \text{ ks/m}^3)(9.8 \text{ m/s}^2) - (600 \text{ kg/m}^3)(9.8 \text{ m/s}^2) = (600 \text{ kg/m}^3)a$

$a = 6.5 \text{ m/s}^2$ upward

This is only the initial acceleration!
Once the wood starts to move it will begin to feel a drag/resistance force.

30. (A) $\dfrac{\Delta V}{\Delta t} = Av = \pi r^2 v = 0.0075 \; \dfrac{\text{m}^3}{\text{s}}$. Remember to convert your units!
    (B) The volume of the pool is 216 m³. Divide the volume of the pool by the volume flow rate, and you get 28,800 seconds to fill the pool, which is 8 hours.

31. (A) Conservation of mass dictates that the mass of water flowing through the river must also flow through the canyon. Since water is incompressible, the volume flow rate remains constant.
    (B) The continuity equation, which is based on the concept of conservation of mass, tells us that $A_1 v_1 = A_2 v_2$. Therefore, the velocity of the water increases in the canyon because there is less flow area for the same volume of mass to move through.

32. First, the ball is held up by the increased pressure from the air hitting its bottom surface. This supplies the upward force that balances gravity. The pressure in a moving fluid is less than in a static fluid. When the ball begins to move out of the stream, the higher pressure of the static air pushes the ball back into the lower pressure airstream.

33. The bottom of the wing is flat, and the air moves past it without changing speed. The top of the wing arches upward. This reduces the area through which the air can move. Therefore, the air must speed up as it passes over the top of the wing (continuity equation). The faster air has a lower pressure (Bernoulli's equation). This creates a pressure differential between the top and bottom of the wing that creates a lifting force.

Questions 34 and 35

34. Bernoulli's equation tells us that

$(P + \rho g y + \tfrac{1}{2}\rho v^2)_{\text{at the top of the fluid}} = (P + \rho g y + \tfrac{1}{2}\rho v^2)_{\text{at the opening}}$

These steps can be taken to simplify the equation:
- Since both the top of the fluid and the opening where the fluid flows out are open to the atmosphere, the pressure is the same in both places (1 atm). Thus, the pressure on both sides of the equation will cancel out.
- The velocity of the fluid at the top is very small and is assumed to be zero.
- At the opening (spigot), $y = 0$.
- At the top of the fluid, $y = h$.

This leaves us with this equation:

$$(P + \rho g h + \tfrac{1}{2}\rho v^2)_{\text{at the top of the fluid}} = (P + \rho g y + \tfrac{1}{2}\rho v^2)_{\text{at the opening}}$$

$$\rho g h_{\text{at the top of the fluid}} = \tfrac{1}{2}\rho v^2_{\text{at the opening}}$$

Next we can cancel out the densities because we are assuming the fluid is incompressible:

$$\rho g h_{\text{at the top of the fluid}} = \tfrac{1}{2}\rho v^2_{\text{at the opening}}$$

$$g h_{\text{at the top of the fluid}} = \tfrac{1}{2}v^2_{\text{at the opening}}$$

Solving for the velocity at the opening:

$$v_{\text{at the opening}} = \sqrt{2gh}_{\text{at the top of the fluid}}.$$

35.

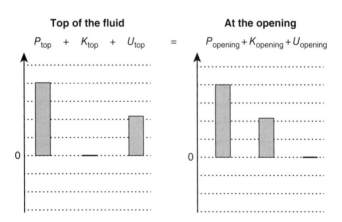

Questions 36 and 37

36. The gravitational potential energy at both ends of the tube are the same as they have the same average height. So we can choose to set the gravitational potential energy at whatever we want because it will not change from side 1 to side 2. (For simplicity, I have chosen it to be zero.) The kinetic energy increases due to continuity from side 1 to side 2. This means the pressure inside the fluid must decrease by the same amount that the kinetic energy increases.

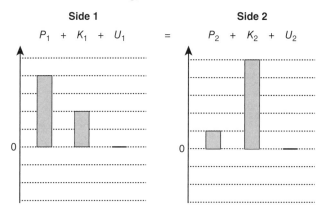

37. The pressure decreases in the fluid. Therefore, the size of the bubbles increases moving from side 1 to side 2.

Questions 38 and 39

38. The water slows down because the cross-sectional area of the 6-cm pipe is larger than that of the two 3-cm pipes.

39. Remember to convert units, and find the radius of each pipe. Also, two pipes lead into a single pipe:

$$A_1 v_1 + A_1 v_1 = A_2 v_2$$
$$2(\pi r_1^2 v_1) = \pi r_2^2 v_2$$
$$2 r_1^2 v_1 = r_2^2 v_2$$
$$v_2 = \frac{2 r_1^2 v_1}{r_2^2} = 4 \text{ m/s}$$

40. (A) Continuity tells us that the velocity of the water is fastest at location 1 and is uniformly slower at locations 2, 3, and 4.

(B) There is a lot going on here.
- Continuity tells us that the speed is slower at locations 2, 3, and 4.
- Bernoulli tells us that the slower the fluid moves, the higher the pressure. This means the pressure at location 2 is higher than at location 1.
- Static pressure tells us that the lower the pipe is, the higher the pressure. This means that in the large pipe, location 3 has the highest pressure and location 4 has the lowest.

Combining all these considerations, we can see that the pressure is highest at location 3 followed by location 2. It is impossible to compare locations 1 and 4 accurately without more information about the pipe diameters. We just know the pressures at location 1 and 4 are lower than at location 2.

41. The slower the water, the higher the pressure. Therefore, due to continuity, nozzle 1 will squirt the water highest, followed by location 3, and finally location 2.

42. Using Bernoulli's equation to connect the surface of the lake to the exit, we see that the velocity of the lake is zero, and the height of the exit is also zero:

$$\left(P + \rho g y + \frac{1}{2}\rho v^2\right)_{lake} = \left(P + \rho g y + \frac{1}{2}\rho v^2\right)_{exit}$$

$$(P + \rho g y)_{lake} = \left(P + \frac{1}{2}\rho v^2\right)_{exit}$$

The top of the lake and the exit are both open to the atmosphere. Thus, the pressure at the top of the lake and at the exit are both atmospheric. Therefore P cancels out.

$$(\rho g y)_{lake} = \left(\frac{1}{2}\rho v^2\right)_{exit}$$

The density of the water is constant.

$$(g y)_{lake} = \left(\frac{1}{2} v^2\right)_{exit}$$

The surface of the lake is 30 m above the exit.

$$v_{exit} = \sqrt{2 g y_{lake}} = 24 \text{ m/s}$$

AP-Style Multiple-Choice Questions

43. (B) The weight of the blocks is balanced by the buoyancy force. Since the blocks are half submerged, their densities are half that of water:

$$F_g = F_B$$
$$(mg)_{block} = (\rho V g)_{water}$$
$$(\rho V g)_{block} = (\rho V g)_{water}$$
$$(\rho V)_{block} = \left(\rho\left(\frac{1}{2}V_{block}\right)\right)_{water}$$
$$\rho_{block} = \frac{1}{2}\rho_{water}$$

44. (D) The exit velocity is proportional to the water pressure at the hole, which is proportional to the depth of the water: $P = P_0 + \rho g h$.

Questions 45 and 46

45. (A) They are all the same because the heights are the same: $P = P_0 + \rho g h$.

46. (D) The pressure is the same at the bottom of each container.

The area of the bottom surface $= \pi \left(\dfrac{d}{2}\right)^2$

$$F = PA = P\pi \left(\dfrac{d}{2}\right)^2$$

Inserting the diameter of each container, we see that the force on the smaller container is 1/16 that of the larger container.

47. (A) The pressure at the top of both pistons must be the same:

$$\left(\dfrac{F}{A}\right)_{\text{left}} = \left(\dfrac{F}{A}\right)_{\text{right}}$$

$$\left(\dfrac{mg}{\pi r^2}\right)_{\text{left}} = \left(\dfrac{mg}{\pi r^2}\right)_{\text{right}}$$

$$\dfrac{m_{\text{left}}}{r_{\text{left}}^2} = \dfrac{m_{\text{right}}}{r_{\text{right}}^2} = \dfrac{m_{\text{right}}}{(4r_{\text{left}})^2}$$

$$16 m_{\text{left}} = m_{\text{right}}$$

48. (B and D) The sum of the forces must equal zero. Therefore, the buoyancy force upward must equal gravity downward minus the tension upward. Buoyancy force is $\rho V g$, and the volume of the block is $a^2(h_1 - h_2)$.

49. (B) All the blocks are floating. Therefore, buoyancy force must equal gravity force on each. A and B weigh the most. C weighs less.

50. (A) Both blocks are floating in equilibrium. Therefore, the net force on each must be zero.

51. (C) The cube on the bottom has the greatest mass and a density greater than water. The floating, submerged cubes have the same density as water. The floating, partially submerged cube has a density less than that of water.

52. (B) The block has a mass of 190 g, but the reading on the spring scale is only 120 g when submerged in the water. This means that the buoyancy force equals 70 g. Newton's third law reminds us that this force is upward on the block and downward on the water which increases the balance reading. When the metal block is lifted out of the water, the buoyancy force between the block and the water disappears. This makes the spring scale reading go up by 70 g and the balance reading go down by 70 g.

53. (A) The speed increases as it passes through the region of plaque, and the pressure in the fluid decreases. Thus, the cell will increase in size.

54. (D) Using the conservation of mass/continuity equation, we see that the water must be slower at the hydrant than at the nozzle: $A_1v_1 = A_2v_2$. The area of the hose is proportional to the diameter squared: $\pi r_1^2 v_1 = \pi r_2^2 v_2$. This gives us a velocity in the hose of 1.125 m/s.

Using conservation of energy/Bernoulli's equation and assuming the exit pressure is atmospheric,

$$\left(P + \rho g y + \frac{1}{2}\rho v^2\right)_1 = \left(P + \rho g y + \frac{1}{2}\rho v^2\right)_2$$

$$\left(P + 0 + \frac{1}{2}(1{,}000 \text{ kg/m}^3)(1.125 \text{ m/s})^2\right)_1$$

$$= \left(100{,}000 \text{ Pa} + (1{,}000 \text{ kg/m}^3)(10 \text{ m/s}^2)(6 \text{ m}) + \frac{1}{2}(1{,}000 \text{ kg/m}^3)(18 \text{ m/s})^2\right)_2$$

Plugging in our values, we get a required pressure of 3.2×10^5 Pa at the hydrant.

55. (C) Be careful to get all your units straight before you start. Using the conservation of mass/continuity equation, $A_1v_1 = A_2v_2$. The area is proportional to the diameter squared, and there are twenty openings in the showerhead:

$$\pi r_1^2 v_1 = 20(\pi r_2)^2 v_2$$

This gives us a velocity at the pipe of 5 v.

AP-Style Free-Response Questions

56. (a) The bubble's size will increase as it moves upward because the external pressure from the water decreases as the bubble rises. The amount of air inside the bubble remains the same during ascent. As the bubble increases in volume, the buoyancy force increases, but the gravity force on the bubble remains the same. Therefore, the bubble accelerates upward at an increasing rate as it ascends.

(b) Buoyancy force (upward) would be larger than the gravity force (downward). Notice that the pressure on the bubble is not drawn because it is not a force. The net effect of the pressure is the buoyancy force itself.

(c) The gas atoms collide with and bounce off the water molecules, changing the gas atoms' momentum. This imparts a force on the water that keeps the bubble from collapsing.

(d) i. $P_D = P_S + \rho g D$

ii. If the temperature remains constant, then the pressure times volume will remain constant:

$$P_S V_S = P_D V_D = (P_S + \rho g D) V_D$$
$$V_S = \frac{(P_S + \rho g D) V_D}{P_S}$$

(e) i. The sketch should have the shape of an adiabatic process, with an upward concave curve similar to that for an inverse relationship. The initial and ending points should be marked. See figure.

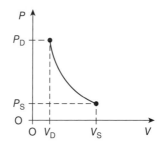

ii. The $P_S V_S$ is less than the $P_D V_D$ value. Since there is no heat transfer between the water and the bubble, the process is an adiabatic expansion. This will cause both the temperature and the PV value to decrease.

57. (a) No, the beaker will not fill any faster. Due to conservation of mass, the mass flow rate (or volume flow rate) must remain the same if the nozzle is removed.

(b) Using the continuity equation,

$$A_1 v_1 = A_2 v_2$$
$$\pi r_1^2 v_1 = \pi r_2^2 v_2$$

Since the radius is cut in half, the exit velocity will be four times larger than in the hose: $v_2 = 1.6$ m/s.

(c) The volume flow rate is

$$A_1 v_1 = \frac{V}{t}$$
$$\pi r_1^2 v_1 = \frac{V}{t}$$

Remember to convert your units! The final time rounded to two significant digits is 8.0 seconds.

(d) $\left(p + \rho g y + \frac{1}{2}\rho v^2 \right)_A = \left(p + \rho g y + \frac{1}{2}\rho v^2 \right)_{nozzle}$

Bernoulli's equation tells us that the $\rho g y$ term and the $\frac{1}{2}\rho v^2$ terms will both be smaller for point A since point A is lower and has a slower oil velocity than at the exit nozzle. Therefore, the pressure at point A must be greater than at the nozzle.

(e) The net force from the oil will be greater on the bottom of the 1,000-ml beaker than on the bottom of the 200-ml graduated cylinder. The pressure at the bottom of both is the same because the height of the oil is the same. However, the beaker must have a larger area base. Therefore, the net force is greater on the bottom of the beaker. Another way to think about it is to say that there is more oil mass in the beaker, so there must be more oil weight for the beaker to support.

(f) i. The acceleration of the cube is zero. Therefore, the forces on the cube cancel:

$$F_{scale} + F_B = F_g$$
$$F_{scale} = F_g - F_B$$
$$F_{scale} = (\rho V g)_{cube} - (\rho V g)_{oil} = (\rho_{cube} - \rho_{oil})Vg = 13 \text{ N}$$

Remember to convert your units! The volume of the cube is 0.000125 m³.

ii. The spring scale reading remains the same. The difference in pressure on the bottom of the cube and the top of the cube does not change once the cube is fully submerged. The buoyancy force also does not change once the cube is completely submerged.

iii. As the cube moves from position 1 to position 2, the pressure on the bottom of the beaker goes up because the oil level rises as the cube displaces fluid volume. This increases the static pressure on the bottom of the beaker. Once the cube is submerged, the oil level no longer rises. Therefore, the pressure on the bottom of the beaker remains the same as the cube is lowered from position 2 to position 3.

Chapter 2: Thermodynamics and Gases

Skill-Building Questions

58. Differences:
- Molecular behavior: In gases, the molecules are not connected to each other and are far apart compared to their size. Each molecule flies around freely. The molecules only interact with each other when they collide. In liquids, the molecules are very close together and exert bonding forces on each other. The bonding is strong enough to hold the fluid together but is not strong enough to hold a definite shape like in a solid.
- Volume and shape: Gases expand to fill the volume of their container. They do not have a defined surface. Liquids have a definite volume and a defined surface.
- Compressibility: Due to the empty space between the molecules, the volume of a gas can easily be changed. (Gases are said to be compressible.) Since the molecules are close together in a liquid, it cannot really be compressed. Due to the bonding between the molecules, liquids can't be easily compressed or expanded. (Liquids are incompressible.)

Similarities:
- Both gases and liquids flow.
- Since they flow, neither gases nor liquids have a definite shape.
- Both gases and liquids exert pressure on the containers that confine them.

59. Gas atoms collide with and rebound off of a surface, imparting a tiny impulse on the surface. Moles of atoms continually colliding with the surface impart a constant force over the area of the surface. This creates a constant gas pressure on the surface.

60. The motion of the atoms in a gas is random. Thus, the atoms collide with the surface in no preferred direction. In addition, when a gas atom bounces off a surface in an elastic collision, only the atom's momentum that is perpendicular to the surface is changed. This means the impulse between the surface and the gas atom is also perpendicular to the surface. Thus, a gas always exerts a perpendicular force on the surface with which it is in contact.

61. The suction cup forms an airtight seal with a smooth surface. When the suction cup is pulled away from the surface, the volume inside increases, causing the pressure inside the cup to decrease below the value of the external fluid pressure. This pressure difference creates a net force that pushed the suction cup to the surface even though it is not physically connected to the surface.

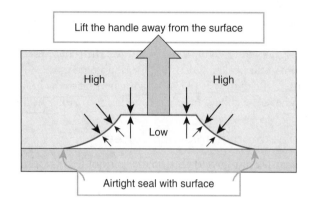

62. A large number of molecules comprise a gas. These molecules are in random motion. All gas energy is in the form of kinetic energy from the moving molecules. The molecules are assumed to be solid, point-like masses, with no appreciable size. They interact with other molecules and surfaces through elastic collisions and obeying Newton's laws.

63. Temperature is a direct measure of the average kinetic energy of the particles that make up the gas: $\bar{K} = \frac{3}{2}k_B T$.

64. As the gas decreases in temperature, the average kinetic energy of the molecules decreases to half the original value: $\bar{K} = \frac{3}{2}k_B T$.

65. The velocity of the gas molecules is proportional to the square root of the temperature:

$$v_{rms} = \sqrt{\frac{3k_B T}{m_{\text{gas molecule}}}}$$

$$v_{rms} \propto \sqrt{T}$$

Since the final temperature is divided by 2, the final velocity will equal the initial velocity divided by $\sqrt{2}$:

$$v_{final} = \frac{v_{inital}}{\sqrt{2}}$$

66. The peak of Ar must be the highest and the farthest to the left. Argon atoms move the slowest because argon has the highest mass. The argon atoms are bunched together at lower speeds. The area under the curve represents the total number of atoms. Therefore, the peak is higher. By reverse reasoning, the peak of the lighter mass He atom distribution must be the lowest and the farthest to the right. The areas under the three curves should be the same as all three samples have 2 moles of gas.

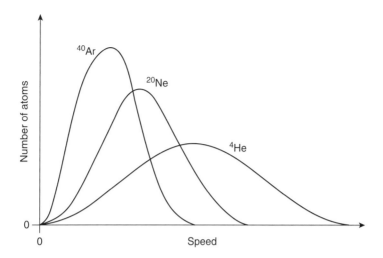

67. **(A)** The nitrogen temperature decreases. The temperature of the carbon monoxide increases. The two gases reach thermal equilibrium at a temperature between 200 K and 400 K, but the final temperature of the combined gas will be closer to 200 K because there are more "cold" CO molecules than "hot" N_2 molecules.

(B) Thermal energy will flow from the hotter nitrogen gas to the colder carbon monoxide. On a macroscopic scale, heat always flows from hot to cold. This is due to the microscopic behavior of faster moving "hot" molecules colliding with, and transferring energy/momentum to, slower moving "cold" molecules.

(C) The carbon monoxide distribution will be shifted to the left of the nitrogen. The CO peak must be higher than that of the nitrogen because there are more molecules of CO, and all the molecules are bunched in a smaller speed distribution. The area under the CO curve should be twice as large as the area under the N_2 curve because there are twice as many molecules.

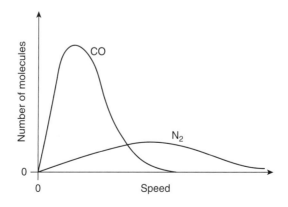

(D) The two gases have the same molecular mass; therefore, they will have the same shaped distribution with the same peak speed. The carbon monoxide distribution should be shifted to the right and have a peak lower than its initial condition because the gas is now spread out over a wider speed distribution. However, the CO peak must still be higher than the N_2 peak because there are more molecules of CO. The nitrogen speed distribution should be higher than its initial value and shifted to the left. The area under the CO curve should be twice as large as the area under the N_2 curve because there are twice as many molecules.

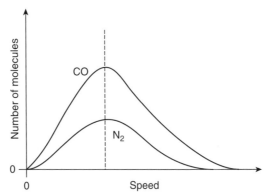

68. There are several different ways to perform this experiment. Here is one example.

(A) Equipment: pressure sensor, syringe with a movable plunger with markings for volume.

(B)

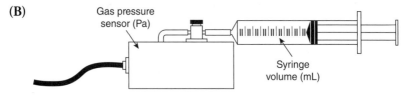

(C) Procedure:
 1. Pull the syringe plunger to a large volume.
 2. Connect the syringe to the pressure sensor.
 3. Record the volume and pressure.
 4. Push the plunger to a smaller volume. Record the new pressure and volume.
 5. Repeat.

(D) The graph should be an inverse relationship curve.

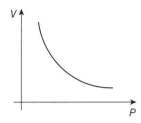

69. There are several different ways to perform this experiment. Here is one example.

(A) Equipment: balloon, ruler or large Vernier calipers, thermometer, incubator or oven, refrigerator, and freezer.

(B)

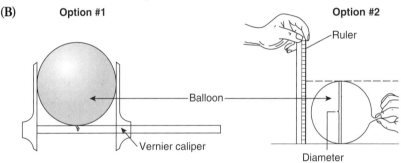

(C) Procedure:
1. Inflate a round balloon. Allow it to come to room temperature.
2. Measure the diameter of the balloon using a large Vernier caliper or a ruler. It may be helpful to take a picture of the balloon next to the ruler and use the picture to determine the diameter of the balloon. Note: This measurement must be done quickly as the air temperature inside the balloon, and size of the balloon, change quickly when handled.
3. Repeat this process with hot/cold air inside an incubator or oven, refrigerator, and freezer. Be sure to allow time for the air in the balloon to come to thermal equilibrium with the external environment in each case.
4. Use the diameter measurement to calculate the volume of the balloon.

(D) Extrapolate the data to a zero volume. The temperature at $V = 0$ will be our experimental value for absolute zero.

(E) The graph should be linear. The *x*-intercept will be our experimental prediction for absolute zero.

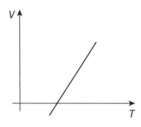

Questions 70–73

70. (C) The graph shows pressure as a function of volume. Therefore, temperature and number of moles must remain constant. This eliminates experiments A and D, which show a temperature change. Experiment B indicates a change in the amount of gas. Experiment C shows a movable piston to change the volume and pressure at a constant temperature and number of moles.

71. (A) The graph indicates a change in volume as a function of temperature. Both experiments A and D indicate a change in temperature, but only the apparatus in experiment A can have a change in volume.

72. (C) Experiment D shows addition of heat without any change in volume or moles of gas. Thus, the temperature and pressure must be changing.

73. (B) The Ideal Gas model indicates a direct relationship between pressure and the number of gas molecules: $PV = nRT$.

74. Convert the temperature to Kelvin, and use the Ideal Gas model to get approximately $n = 1.03$ mol of gas and 6.2×10^{23} molecules. Remember that these are experimental data. Using different data pairs will yield slightly different answers due to variations in the data. You can improve your accuracy by using each data pair to calculate the number of moles and then average them. Or, better yet, graph V as a function of $(1/P)$ and use the slope of the best fit line to determine the number of moles.

75. The average kinetic energy of gas molecules is: $\bar{K} = \frac{3}{2} k_B T$. If we simply multiply by the number of gas molecules (N), we will arrive at the total thermal energy of the gas. We could also use Avogadro's number (N_A) and the number of moles (n) to determine the thermal energy of the gas:

$$U_{\text{Thermal}} = N\left(\frac{3}{2} k_B T\right) = nN_A\left(\frac{3}{2} k_B T\right) = \frac{3}{2} nRT$$

Since an ideal gas has no potential energy, the internal energy of the gas is equal to the thermal energy of the gas.

76. Internal energy is the sum of the kinetic and potential energies of all the atoms inside the material. Ideal gas atoms are assumed not to interact with each other at a distance. Therefore, there is no potential energy bonding between the atoms to contribute to the overall internal energy. Note that both liquids and solids have internal atomic potential energy as a component of their internal energy because their atoms are more closely packed together than in a gas.

77. **(A)** Due to gas pressure, the gas pushes outward on the piston. The force from the piston is inward on the gas. When the gas is being compressed, the force from the piston and the displacement of the piston are in the same direction causing the work done on the gas to be positive. When the gas expands, the piston is still pushing inward on the gas but the piston is moving in the opposite direction. This causes the work done on the gas to be negative. This means that the work done on the gas is negative when the gas expands in volume. Thus the equation for work done on the gas is: $W = -P\Delta V$

(B) As the piston moves downward, it collides with the gas molecules, imparting energy to the molecules by collision. This causes the molecules to move faster, which increases both the temperature and the thermal energy of the gas.

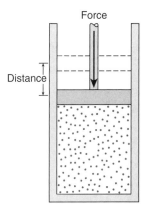

(C) When a gas expands and moves a piston, the gas atoms are colliding with the piston, imparting a force through a distance. This causes the gas molecules to slow down as they lose kinetic energy in the collision, thus decreasing the internal energy of the gas.

78. There are several different ways to perform this experiment. Here is one example.

(A) Equipment: Syringe with volume markings mounted vertically with a platform on top of which masses can be added.

(B) Load masses on the top platform. Measure the volume.

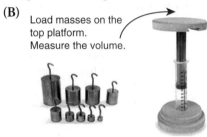

(C) Procedure:
1. Measure/calculate the cross-sectional area (A) of the syringe/plunger.
2. Place the plunger in the syringe. Record the starting volume and atmospheric pressure. Seal the front of the syringe and mount it vertically. Attach the top platform.
3. Place a mass on top of the platform. Allow the plunger to descend and reach equilibrium.
4. Record the new volume and calculate the change in volume due to the mass being added to the platform: $\Delta V = V_{original} - V_{new}$.
5. Calculate the net pressure due to the mass: ($P_{net} = mg/A$).
6. Calculate the work done by the mass: $W = P_{net}(\Delta V)$.
7. Add another mass to the first and repeat steps 3–6.
8. Sum up all the incremental works from each added mass to find the total work done by all of the combined masses.

(D)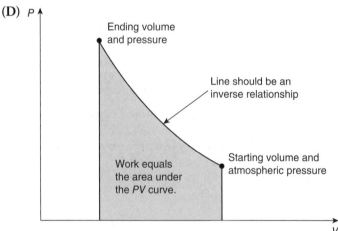

79. The energy of a system can be changed by doing work on it or by adding heat to it. Work is a process where a force moves the system through a distance that will change the energy of the system. Heat is the process of directly transferring thermal energy from one system to another.

80. Conduction, convection, and radiation.
- Conduction is the process by which faster moving "hot" atoms collide with slower moving "cold" atoms and transfer kinetic energy by collision from atom to adjacent atom. By these collisions, thermal energy is transferred from the hotter area to a colder area through objects that are in contact.
- Convection: In a hot fluid, the atoms vibrate faster, on average, and take up more space than do those in colder fluids. This causes the hot fluid to be less dense. The less dense, hot fluid is more buoyant and will rise upward in the opposite direction of gravity.
- Radiation is due to the vibration of charged particles inside the atom. Vibrating charged particles generate electromagnetic waves (photons of light) that radiate away from the object. Hotter objects, on average, have faster moving atoms and emit higher energy photons. Colder objects will emit lower energy photons. Only at absolute zero would an object not radiate any electromagnetic radiation.

Questions 81–83

There are several methods to perform these experiments. One possible method is listed.

81. (A) The longest copper rod, multiple temperature sensors, ice, a beaker, water, and clamps and stands to hold the copper rod

  (B) Procedure:
   1. Attach the temperature sensors at regular intervals along the copper rod.
   2. Record the initial temperature of all sensors.
   3. Fill the beaker with water and ice.
   4. Using the clamps and stands, position the copper rod vertically with its lower end in the beaker.
   5. Start a stopwatch.
   6. At 5-minute intervals, read the temperatures along the length of the copper rod.
   7. The temperature changes in the sensors will indicate the relationship between length and the thermal conductivity rate.

(C) The conductive heat transfer rate is inversely proportional to the length of the path:

$$\frac{Q}{\Delta t} = \frac{kA\Delta T}{L}$$

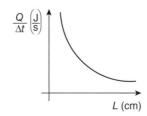

82. (A) Any one rod, two temperature sensors, ice, a beaker, water, a hot plate, and clamps and stands to hold the rod
    (B) Procedure:
      1. Attach one temperature sensor to the end of the rod.
      2. Record the initial temperature, which should be room temperature. If not, allow the rod to come to room temperature.
      3. Fill the beaker with water. Place on the hot plate, which is turned to the lowest setting. Use the second sensor to record the temperature of the water.
      4. When the water temperature has reached a stable equilibrium temperature, use the clamps and stands to position the rod vertically with its lower end in the beaker. The temperature probe end should be on top, not in the water!
      5. Start a stopwatch.
      6. At 10 minutes, record the temperature at the top end of the rod. Calculate the delta temperature of the end of the rod over the 10-minute interval.
      7. Remove the rod from the water and allow it to cool to room temperature. Verify this with the temperature probe.
      8. Repeat steps 3–6 for multiple hot plate and water temperature settings.
      9. The temperature change of the end of the rod for multiple water temperatures will indicate the relationship between temperature difference between the ends of the rod and thermal conductivity rate.

(C) The conductive heat transfer rate should be directly proportional to the difference in temperature between the ends of the rod:

$$\frac{Q}{\Delta t} = \frac{kA\Delta T}{L}$$

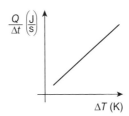

83. (A) The four steel rods of varying thickness, wax, a Bunsen burner, and clamps and stands to hold the steel rods
    (B) Procedure:
        1. Using the clamps and stands, position the steel rod horizontally so that one end can be inserted into the flame of the Bunsen burner.
        2. Cut four small, identical pieces of wax. Place a piece of wax at one end of the steel rod. Place the other end in the center of the Bunsen burner flame.
        3. Time how long it takes the wax to melt completely.
        4. Repeat this process for the other three rods, which have different cross-sectional areas.
        5. Since it should take the same amount of heat to melt each piece of wax, the shorter the time to melt, the higher is the heat transfer rate.
    (C) The heat transfer rate should show a direct relationship with the cross-sectional area of the rod. So the larger the cross-sectional area of the rod, the shorter the time it should take to melt the wax:

$$\frac{Q}{\Delta t} = \frac{kA\Delta T}{L}$$

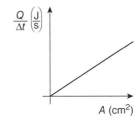

84. **(A)** The atoms in the coffee have a higher average kinetic energy than those of the spoon. Thus, when the atoms of the two interact by collision, it is statistically more likely that thermal energy will transfer from the more energetic "coffee" atoms to the less energetic "spoon" atoms. Summed up over the vast number of atoms interacting, the net effect is always that heat flows from the hotter to the colder object.

   (B) When objects have the same temperature, the atoms that comprise the objects have identical average kinetic energies. Thus, when the atoms of the two objects interact by collision, thermal energy transfers back and forth between the objects at statistically identical rates. Therefore, the net heat transfer between two objects of the same temperature is always zero. We call this thermal equilibrium.

   (C) Both the spoon and the coffee have identical **average** kinetic energies. However, the question implies that the coffee is much larger (has more atoms). Therefore, the coffee will have more total thermal energy simply because it has more atoms.

   (D) The spoon gains heat, so its entropy rises. The coffee loses heat, so its entropy decreases. This is an irreversible process; that is, the spoon and coffee will not naturally return to their origin states. Heat will not naturally flow back from the spoon into the coffee to return both to their original "hot" and "cold" temperatures. In all irreversible processes, the entropy of the system rises. Therefore, the rise in entropy of the spoon is greater than the loss of entropy of the coffee.

85. In theory, a reversible process is one in which there is no entropy rise. Thus, the process can move in either direction, forward or backward, and nature has no preference to the direction. For systems made up of atoms, the reality is that every process involving the transfer of energy from one form to another or from one object to another is irreversible. This is because at least some organized energy of the system is lost to thermal energy. This is a natural process that occurs in only one direction. It is impossible for the lost thermal energy to return to its original organized state without the input of work into the system.

Questions 86–89

	PV and energy explanations	PV diagram	Energy bar chart
86.	This is an isochoric process with no volume change. The temperature decreases to one-third of the original value. Therefore, the pressure must decrease to one-third of the original as well because $P \propto T$. Heat is removed from the gas. Work is zero, and the internal energy decreases.	$P\ (\times 10^5\ \text{Pa})$ vs $V\ (\times 10^{-4}\ \text{m}^3)$: point at (6, 3) moving down to (6, 1).	$W + Q = \Delta U_{\text{gas}}$; $W = 0$, Q negative, ΔU_{gas} negative.
87.	The mass on the top of the piston remains constant, making this is an isobaric process. The temperature increases to twice its original value, so the volume must also double because $V \propto T$. Heat is added to the gas. Negative work is done as the gas expands. The temperature of the gas goes up, increasing the internal energy.	$P\ (\times 10^5\ \text{Pa})$ vs $V\ (\times 10^{-4}\ \text{m}^3)$: point at (6, 3) moving up to (12, 3).	$W + Q = \Delta U_{\text{gas}}$; W negative, Q positive, ΔU_{gas} positive.

	PV and energy explanations	PV diagram	Energy bar chart
88.	The gas stays in thermal equilibrium with the environment, making this an isothermal process. The volume decreases to one-third the initial value. The pressure must triple because $P \propto \dfrac{1}{V}$. Note that the process will pass through all points that have the same PV value. An additional data point on the path is shown in the figure. The gas is compressed, making the work positive. The internal energy remains constant because the temperature is constant. Heat flows out of the gas, keeping the gas at the same temperature.	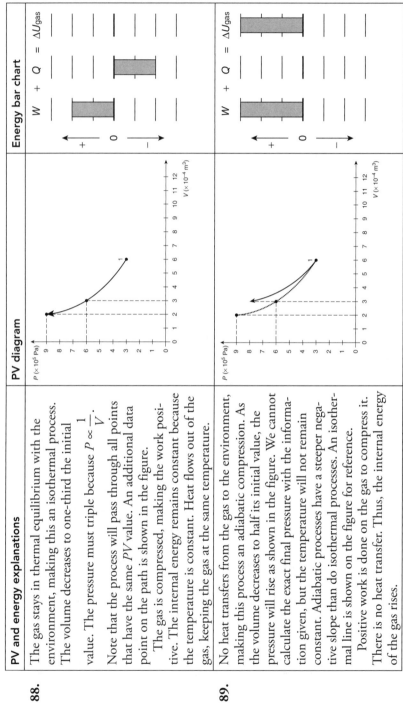	
89.	No heat transfers from the gas to the environment, making this process an adiabatic compression. As the volume decreases to half its initial value, the pressure will rise as shown in the figure. We cannot calculate the exact final pressure with the information given, but the temperature will not remain constant. Adiabatic processes have a steeper negative slope than do isothermal processes. An isothermal line is shown on the figure for reference. Positive work is done on the gas to compress it. There is no heat transfer. Thus, the internal energy of the gas rises.		

90. (A) Key features of the graph:
- All the blocks are identical and will reach equilibrium at the central temperature of 50°C.
- The temperature of block C continuously rises to equilibrium.
- Block A gains energy from block B, making block A initially rise above the equilibrium temperature of 50°C.
- Block B loses energy to both blocks, taking it below the temperature of block A. Block A then begins returning energy to block B. The sample solution shows that the temperature of block B drops below 50°C. This may or may not occur. The critical feature of the graph is that the temperature of block B drops below that of block A but remains above that of block C.

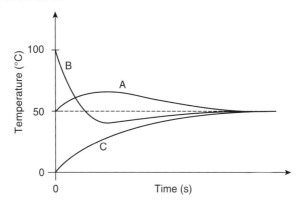

(B) The entropy rises as heat flows into block C. This expands the velocity and kinetic energy profile of the atoms. This, in turn, increases the overall disorder of the block as there are more thermal microstates for the atoms to occupy.

(C) This is an irreversible process. Therefore, the entropy of the system increases.

91. This is not an isothermal process. The end points of the curve do not have the same $P \times V$ value.

92. Temperature is directly proportional to the product of the pressure and volume (PV). The highest PV value is point D, so it has the highest temperature. Point B has the lowest PV and thus the lowest temperature.

93. The temperature of the gas cannot be determined without knowing the amount of gas present. We need to know the number of moles of gas.

94. (A) Sketch any path that ends at a volume of 2V but a lower pressure P.

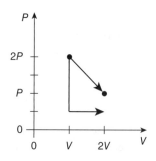

(B) The work done in the original process is equal to the area under the curve. The trapezoidal area under the curve is 1.5PV. The new process has the same area under the curve.

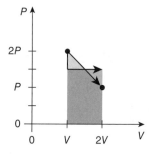

(C) There are many possible solutions, but the path must stay above the original line. The new process has a higher final temperature and, therefore, a positive ΔU. The work is a larger negative value than the original. The only possible way for this to happen is for more heat to be added to the gas in the new process.

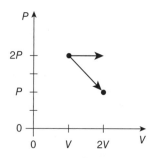

95. From point A to point B, the temperature of the gas increases while negative work is being done. This can only occur when heat is added to the gas. Therefore, the entropy of the gas is increasing. From point B to point C, the temperature drops while no work is being done. Therefore, heat is leaving the gas, and the entropy is decreasing.

96.

Name of process	ΔT	ΔU	W	Q	Notes
Adiabatic	−	−	−	0	ΔU and W are the same value.
Isothermal	0	0	+	−	Q and W are the same magnitude.
Isochoric (isovolumetric)	+	+	0	+	ΔU and Q are the same value.
Isobaric	−	−	+	−	Q has a larger magnitude than W.

ΔT is found by seeing how the path moves through PV values using the Ideal Gas model: $\Delta(PV) = nR\Delta T$.

ΔT and ΔU always have the same sign: $\Delta U_{\text{Thermal}} = N\left(\frac{3}{2}k_B\Delta T\right) = \frac{3}{2}nR\Delta T$.

Moving to the right is negative work. Moving to the left is positive work. Moving up or down, the volume does not change, and the work is zero: $W = -P\Delta V$.

After finding ΔU and W, use the first law of thermodynamics to calculate Q: $\Delta U = Q + W$.

97. (A) $W = -P\Delta V$ therefore, $W_{CA} > W_{BC} = 0 > W_{AB}$

(B) $\Delta(PV) = nR\Delta T$ therefore, $\Delta T_{AB} > \Delta T_{CA} = 0 > \Delta T_{BC}$

(C) $U_{\text{Thermal}} = N\left(\frac{3}{2}k_BT\right) = \frac{3}{2}nRT$ therefore, $U_B > U_A = U_C$

(D) $T = \dfrac{PV}{nR} = \dfrac{(10 \times 10^5 \text{ Pa})(10 \times 10^{-3} \text{ m}^3)}{(2 \text{ mol})\left(8.31 \dfrac{\text{J}}{\text{mol K}}\right)} = 600 \text{ K}$

(E) The process returns to the original state: $\Delta T = 0$.

(F) Work is the area under the path: 3,750 J.

(G) Since $\Delta U_{CA} = 0, Q = -W = -3{,}750$ J.

Questions 98–100

98. Isochoric—also called a "constant volume process" and occasionally "isovolumetric."

99. Isobaric—constant pressure process.

100. Isothermal—the PV value does not change, so the temperature does not change either.

AP-Style Multiple-Choice Questions

101. (B) Gases with the same temperature will have the same average molecular kinetic energy: $\bar{K} = \frac{3}{2}k_B T$.

102. (A) $v_{rms} = \sqrt{\frac{3k_B T}{m_{gas\ molecule}}}$. Nitrogen is lighter than oxygen. Since the molecular velocity is inversely proportional to the mass of the gas molecules, the nitrogen in the air will have a higher overall speed.

103. (D) The data in the table suggest that gas pressure and volume are inversely related: $P \propto \frac{1}{V}$. This is also seen in the Ideal Gas model: $P = \frac{nRT}{V}$. Therefore, plotting P on the vertical axis and $1/V$ on the horizontal axis will produce a straight line from which the number of moles could be calculated knowing that the slope will equal nRT.

104. (A and D) The Ideal Gas model $\left(P = \frac{nRT}{V}\right)$ shows us that gas pressure is directly related to gas temperature ($P \propto T$). This is seen in the straight line data represented in the graph. The slope of this graph will be equal to $\frac{nR}{V}$. Since the volume of the gas is given, the number of moles—and thus, the number of atoms—in the gas can be calculated using the slope. The x-intercept represents the temperature of the gas when the volume reaches zero. This point is absolute zero. The area of a PV diagram would represent work. We cannot calculate the force from the pressure because we do not know the surface area of the container. We are only given the volume.

105. (B) $PV = nRT$. With the door open between the identical rooms, both the pressure and the volume of the rooms will be the same. This leaves us with the proportion: $n \propto \frac{1}{T}$. Therefore, the colder room will have more gas molecules. In practical terms, colder air has slower moving molecules that will individually occupy less space, making the cold air more dense.

106. **(B)** The number of moles of gas is constant. From the Ideal Gas model $\left(V = \dfrac{nRT}{P}\right)$, we can see that the volume is directly related to temperature and inversely related to the pressure. As the pressure increases, moving from mountain to sea level, the volume will decrease. The temperature increase will actually try to increase the volume of the bottle. We must assume that the pressure change is greater than the temperature change. The concentration of oxygen has no effect on the Ideal Gas behavior of the bottle.

107. **(C)** The ranking will be based on the $P \times V$ value: $T_C > T_B = T_D > T_A$.

108. **(B)** Both paths start and end at the same point. Therefore, the initial and final temperatures are the same, as are the initial and final thermal energies. Process 1 has a higher average pressure for the same volume change. Another way to think about it is to compare the area under the curves. Graph 1 has more area and, therefore, a larger magnitude of work.

109. **(C)** In process 3, the work done is zero: $W = 0$. The temperature and the internal kinetic energy of the gas are increasing. Therefore, thermal energy must be entering the gas: $\Delta U = Q$.

110. **(C)** $K_{\text{average of the gas}} = \dfrac{3}{2} k_B T$ and $K = \dfrac{1}{2} mv^2$. The helium is moving at half the speed and has four times the mass of hydrogen. This means they both have the same average kinetic energies and the same temperatures:

$$K_{\text{helium}} = \dfrac{1}{2}(4m_{\text{hydrogen}})\left(\dfrac{1}{2}v_{\text{hydrogen}}\right)^2 = \dfrac{1}{2}m_{\text{hydrogen}}v_{\text{hydrogen}}^2 = K_{\text{hydrogen}}$$

Objects with the same temperature are in thermal equilibrium and do not transfer any net thermal energy between them.

Questions 111 and 112

111. **(B and C)** The processes end at the same PV value as they started. Thus both processes have the same temperature and thermal energy change: $\Delta U = (3/2)nR\Delta T$.

112. **(A)** In process 1, work is zero, and the internal energy of the gas decreases; therefore, heat leaves the gas. In processes 2 and 3, the internal energy of the gas remains constant, and negative work is done; therefore, heat must enter the gas. In process 4, negative work is done, and the internal energy of the gas increases, so heat enters the gas.

113. (B) The mass of argon is greater. Thus, at the same temperature, argon has the same kinetic energy but a slower average speed. This shifts the curve to the left. Since the number of moles is the same, the peak of the argon graph must be higher to accommodate the same number of atoms.

114. (B) The work done by the gas is positive because the change in volume is negative: $W = -P\Delta V$. The temperature is decreasing as the PV value decreases: $PV = nRT$.

Questions 115 and 116

115. (C) Adding heat to the gas from the resistor will increase the kinetic energy and temperature of the gas. The kinetic energy of the gas is proportional to the velocity squared:

$$E_{\text{added to gas from resistor}} = \Delta K_{\text{average of the gas}} = \frac{1}{2}mv^2$$

Thus, the average velocity of the gas is proportional to the square root of the energy added to the gas (E).

$$v = \sqrt{(2E/m)}$$

116. (B) The temperature is directly related to the energy supplied to the gas:

$$E = K_{\text{average of the gas}} = \frac{3}{2}k_B T$$

And gas pressure is directly related to the temperature: $PV = nRT$. Therefore, pressure is directly related to the energy (E).

AP-Style Free-Response Questions

117. Note that all the numbers in this problem are rounded to two significant digits because that is the accuracy of the data from the graph.

(a) $PV = nRT$, $T = 480$ K. The temperature of the gas is directly related to the average kinetic energy of the gas molecules.

(b) $F = PA = 2{,}000$ N. The gas molecules collide with the piston in a momentum collision that imparts a tiny force on the piston. The sum of all the individual molecular collision forces is the net force on the piston.

(c) $\Delta U = \frac{3}{2} nR\Delta T$. Since the PV value of the gas decreases, the temperature of the gas decreases in this process. Therefore, the internal energy of the gas also must decrease.

(d) $\Delta U = Q + W$. The internal energy of the gas is decreasing in this process, and the work is positive. Therefore, thermal energy is being removed from the gas. The heat is negative.

(e) The entropy of the gas is decreasing because thermal energy is being removed in this process. This reduces the spread of the speed distribution of the gas, thus reducing disorder.

(f) $\Delta U = \dfrac{3}{2} nR\Delta T$

$= \dfrac{3}{2} nR(T_A - T_C) = \dfrac{3}{2}(1 \text{ mole})\left(8.314 \dfrac{\text{J}}{\text{mole} \cdot K}\right)$
$\times (480 \text{ K} - 120 \text{ K}) = 4{,}500 \text{ J}$

(g) Based on the PV values, the temperature at point A is higher than that at point B. Thus, the peak for A must be at a higher speed than for B. Note that the area under the graphs must be equal because the number of molecules remains the same. This means the peak for A must be lower than that for B.

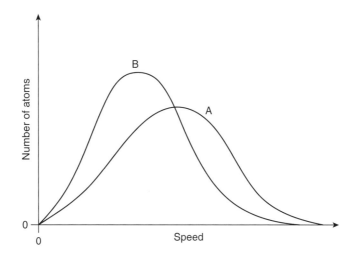

118. There are several different ways to perform this experiment. Here is one example.

(a) Equipment: Beaker filled with water, temperature probe, gas pressure sensor, flask, a stopper with a hole for the pressure sensor, ice, hot plate.

(b)

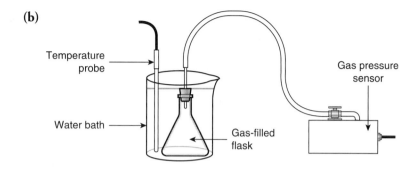

Procedure:
  i. Place the thermometer in the beaker filled with water.
  ii. Attach the gas sensor to the stopper and tightly insert the stopper into the flask. Immerse in the water bath.
  iii. Use ice and the hot plate to adjust the temperature of the water bath. Be sure to allow time to reach equilibrium at each temperature.
  iv. Measure the pressure and temperature at each equilibrium point. Repeat for several equilibrium temperatures.

(c)

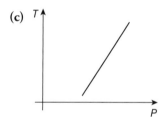

(d) Extrapolate the data to a zero pressure. The temperature at $P = 0$ will be our experimental value for absolute zero.

(e) The set of data from trials 12 through 15 are the best to use because they have a constant temperature and the most data points with the greatest data spread for increased accuracy.

(f) The axis must be labeled with units. Data points must be visible. The best fit curve should be drawn.

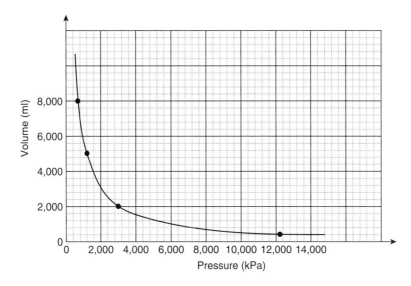

(g) Gas volume is inversely proportional to pressure.

Chapter 3: Electric Force, Field, and Potential

Skill-Building Questions

Questions 119 and 120

119. Polyester has a higher affinity for electrons than for human skin. Due to the frictional contact, electrons will transfer from skin to polyester. The negative polyester and positive skin attract each other.

120. The suspended balloon will be negatively charged by the hair.

(A) The plastic wrap–rubbed glass will be positively charged and will attract the suspended balloon.

(B) The PVC will become negatively charged and will repel the suspended balloon.

(C) The neutral paper will become polarized and will attract the suspended balloon.

121. (A) The negatively charged PVC will transfer some of its charge to the conductive metal pie pan. Likewise, some of the negative charges on the pie pan transfer to the lightweight cereal and are then repelled and fly out of the pan.

(B) The glass rod will be positively charged by the plastic wrap. Held above the cereal, it polarizes the lightweight cereal and attracts it.

(C) Some cereal will fly up and touch the glass rod. The cereal will pick up a net positive charge by transferring electrons to the rod and then fly away again by electrostatic repulsion from the positive rod.

122. Yes! The compass will polarize just like any other material. The north pole of the compass will be polarized positively and the south pole is polarized negatively. The north end of the compass will be attracted to the charged balloon.

123. (A) The clear sticky tape is attracted to the can due to polarization of the can.

(B) The tape is now repelled by the can because the stronger negative charge of the balloon will drive electrons toward the right side of the can, which will repel the tape.

124.

Initial charge of electroscope	Situation	Does the deflection of the leaves increase, decrease, or remain the same?	Justification for your answer
Neutral	Neutral rod touches the metal knob.	Same	Both are neutral. No charge movement when rod touches electroscope.
Neutral	Positively charged rod touches the metal knob.	Increase	Positive rod draws electrons from the electroscope by conduction, leaving the electroscope with a net positive charge.
Neutral	Positively charged rod is brought near but does not touch the metal knob.	Increase	Positive rod attracts electrons to the top of the electroscope, leaving the bottom of the electroscope with a negative charge that separates the leaves by repulsion.
Positive	Neutral rod touches the metal knob.	Decrease	The positively charged electroscope attracts electrons from the rod. Some will transfer to the electroscope by conduction, decreasing the net positive charge of the electroscope.

Positive	Neutral rod is brought near but does not touch the metal knob.	Decrease	The neutral rod will polarize in the presence of the positively charged electroscope. This will cause an attraction between the rod and the electroscope. Some of the net positive charge is attracted to the top of the electroscope, decreasing the net charge at the bottom of the electroscope.
Positive	Negatively charged rod touches the metal knob.	Most likely decrease	The rod and electroscope share charge by conduction. This will neutralize the net charge of the electroscope as electrons move onto it from the rod. (If the negative charge transferred to the electroscope is larger than the original positive charge, the leaves could actually increase in deflection.)
Positive	Negatively charged rod is brought near but does not touch the metal knob.	Decrease	The negative rod will repel electrons in the electroscope downward, making the top of the electroscope more positively charged. Since the net charge of the electroscope remains the same, the bottom end becomes less positively charged, and the leaves do not repel each other as much as before.
Negative	Negatively charged rod touches the metal knob.	It depends…	It depends on which of the two has the larger electric potential. If the rod has a larger negative potential, electrons will flow to the electroscope, and the leaf deflection will increase. If the rod has a smaller negative potential, the reverse will occur.
Negative	Negatively charged rod is brought near but does not touch the metal knob.	Increase	The negative rod will repel electrons in the electroscope, causing the bottom of the electroscope to become even more negatively charged and deflecting the leaves more.

125. **(A)** Two objects with a different affinity for electrons in the outer shell of the atoms will transfer electrons between each other when rubbed together with friction. The friction enables more contact area between the objects and supplies extra energy to facilitate the transfer of electrons between the two materials. One of the objects becomes negatively charged, while the other becomes positive.

(B) When in contact, net charge transfers between two objects until the electric potential is equalized. In metals, this occurs over the entire object. Insulators will only charge at the point of contact. Note that the net charge of the two-object system before and after contact remains the same.

(C) Objects can be charged by induction, where a charged object A is brought close to object B, causing it to become polarized. An escape path is made such that the repelled charge is driven off object B. Finally, the escape path is removed so the charge that was driven off cannot return. Now object B is permanently charged opposite to the charge of object A. See figure.

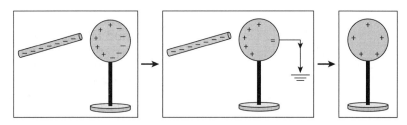

126. **(A)** A red balloon and a blue balloon, thread, kitchen plastic wrap, human hair.

(B) Procedure:
1. Blow up the balloons and tie a long thread to each.
2. Charge the red balloon positively by rubbing the kitchen plastic wrap all over its surface. Charge the blue balloon negatively by rubbing its surface on your hair.
3. Hold each balloon by the thread, and one at a time, bring them close to the charged metal sphere. Observe the results. One balloon should be attracted and the other repelled. The balloon that is repelled will be the same sign charge as the metal sphere.

127. There are several different ways to perform this experiment. Here is one example.

(A) Van de Graaff generator, two conductive spheres on insulating stands, a red balloon and a blue balloon, thread, kitchen plastic wrap, human hair.

(B) Procedure:

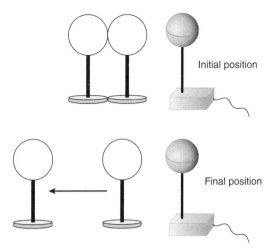

1. Position the conductive spheres and Van de Graaff, as shown in the figure, so both spheres touch. Ground the spheres to make sure they are neutral at the start of the experiment. Turn on the Van de Graaff generator.
2. Move the left sphere away from the right sphere, as shown in the figure. Turn off the generator.
3. Blow up the balloons and tie a long thread to each.
4. Charge the red balloon positively by rubbing the kitchen plastic wrap all over its surface. Charge the blue balloon negatively by rubbing its surface on your hair.
5. Hold each balloon by the thread, and one at a time, bring them close to the two spheres. Observe the results. One balloon should be attracted and the other repelled. The balloon that is repelled will be the same sign charge as the metal sphere.
6. If the student is correct that the charges have been rearranged, one of the spheres should end up positively charged and the other negatively charged due to the influence of the charged Van de Graaff generator.

128. Grounding is a process for removing any excess charge on an object by connecting it to another conductive object of larger size. Usually the larger object is the Earth ("the ground"). When grounded, any excess charge is neutralized by the transfer of electrons between the object and the Earth. Any object electrically connected to the Earth will have an electric potential of zero.

129. (A) Charge added to a conductor migrates to the surface of the object. There will be no excess electric charge inside the object. All excess charge will spread out evenly on the surface of the sphere.

(B) Charge doesn't move through insulators. All the excess charge remains where it was originally placed.

130. **(A)** Both spheres will remain neutral because there is not a conductive pathway for charge to move onto or off of either sphere.

   (B) The conducting system that includes both spheres and the copper rod will become polarized. Sphere A will be positively charged, and sphere B will be negatively charged. The net charge of the system that includes both spheres and the copper rod is still zero because no charge has been added to the system.

   (C) The answer to (A) is now: Sphere A becomes negative by contact, but sphere B remains neutral because wood is an insulator. The answer to (B) is now: Both spheres become negative by contact because there is a conductive pathway connecting both spheres.

131. **(A)** We need to assume that the charges of the balloons, q_1 and q_2, are the same so we can solve the problem. From our free body diagram, we can write the following equations because the forces in the x and y directions must cancel out:

$$T\cos\theta = mg$$

$$T\sin\theta = k\frac{q_1 q_2}{r^2} = k\frac{q^2}{r^2}$$

Solving the first equation for T and substituting into the second equation, we get:

$$mg\frac{\sin\theta}{\cos\theta} = mg\tan\theta = k\frac{q^2}{r^2}$$

$$q = \sqrt{\frac{mgr^2 \tan\theta}{k}} = 2.3 \times 10^{-7} \text{ C}$$

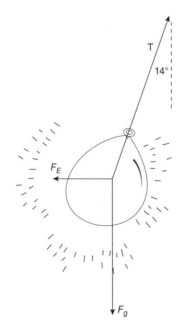

(B) The number of charge carriers can be determined by dividing by the magnitude of the electron charge. Since the balloon was charged by friction with rabbit fur, we know the balloons are negatively charged. Therefore, the excess charge will be electrons.

$$\frac{q}{e} = \frac{2.3 \times 10^{-7} \text{ C}}{\left(1.6 \times 10^{-19} \frac{\text{C}}{\text{elementary charge carrier}}\right)} = 1.5 \times 10^{12} \text{ electrons}$$

(C) As the distance between the balloons decreases, the electric force between them increases. This causes the angle of the thread to increase.

132. (A) All the electric field vectors in the figure point inward toward the charge. Electric field vectors point in the direction of the force on positive charges; therefore, the charge in the figure is negative.

(B) $E_O > E_N = E_P$. The length of the electric field vector indicates the strength of the field.

(C) The electric field is the same magnitude at N and P. Both the electron and the proton have the same magnitude of charge. The electric force = Eq for each is the same. However, the mass of the proton is larger than that of the electron. Therefore, the acceleration of the electron is greater. The electron accelerates in the opposite direction of the field. The proton accelerates in the same direction as the field.

(D) The proton accelerates from rest inward in the direction of the electric field. The acceleration of the proton increases as it moves into a larger electric field closer to the charge. The electron accelerates outward away from the charge. The acceleration decreases as it gets farther away from the charge where the field is weaker. The electron eventually reaches a constant velocity when it is very far away from the charge.

133. (A)

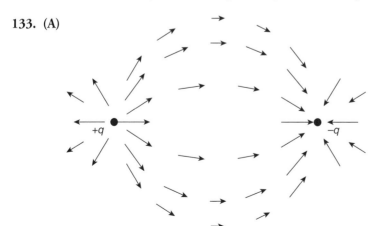

(B)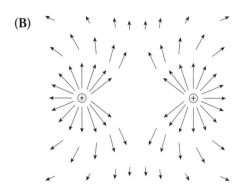

134. Sphere X will have a charge of $-2Q$. Sphere Y has a charge of Q. The negative rod polarizes the three-sphere system, making X negatively charged. When X is moved to the left, it retains this net negative charge, leaving Y and Z with an equal and opposite charge. When the rod is removed, the net positive charge on Y and Z spreads out evenly over both spheres. When Y and Z are separated, each carries half of this positive charge.

135. **(A)** The force on sphere 2 from sphere 1 is $-F$ (to the left) due to Newton's third law.

(B) The force on sphere 3 from sphere 1 is $-\frac{1}{2}F$ because q_2 and r are both doubled: $F_E = k\frac{q_1 q_2}{r^2}$.

(C) $F_2 > F_1 > F_3$. When spheres 1 and 2 touch, nothing happens because they have the same size and charge. (Their electric potentials were already the same.) When sphere 2 touches 3, the final charge of each is $+\frac{1}{2}Q$. Sphere 2 is now forced to the left by both of the other spheres. Sphere 1 is pulled to the right by both of the other spheres. Sphere 3 is repelled to the right by sphere 2 but attracted to sphere 1.

136. **(A)** $F_E = k\frac{q_1 q_2}{r^2} = 8.23 \times 10^{-8}$ N, where the charge of the proton and electron is 1.6×10^{-19} C, and $k = 9 \times 10^9 \,\frac{Nm^2}{C^2}$.

(B) The forces are equal in magnitude but opposite in direction, according to Newton's third law.

(C) The force between the electron and proton is the same. However, the mass of the electron is on the order of 10^{-31} kg, while the proton has a mass on the order of 10^{-27} kg. Therefore, the acceleration of the electron will be on the order of 10^4 times larger than that of the proton: $a = \frac{\Sigma F}{m}$.

(D) The universal gravitational constant is very small (10^{-11}), and the mass of the electron and proton are also very small (10^{-31} kg and 10^{-27} kg, respectively). These values are much smaller than the magnitude of the numbers in the electrostatic equation. This makes the gravitational force negligible compared to the electric force in an atom. Gravity can be ignored.

137. (A) $E = k\dfrac{q}{r^2}$. For the electric field to be zero, the electric fields must be in opposite directions and the same magnitude. This can only occur to the left of the 2-µC charge. Since the −8-µC charge is four times bigger than the 2-µC charge, we need the radius to be twice as far from the −8-µC charge. This occurs at the location of −4 cm.

(B) $F_E = k\dfrac{q_1 q_2}{r^2}$. For the forces to cancel, they must be equal and opposite. Since k and q_1 are the same when calculating the force, the ratio of $\dfrac{q_2}{r^2}$ must be the same: $\dfrac{-8\ \mu C}{(4\ cm)^2} = \dfrac{-4\ \mu C}{r^2}$, so $r = 2.83$ cm. But we need the −4-µC charge to be to the left of the 2-µC charge so the forces are in the opposite direction. Therefore, the location on the x-axis will be −2.83 cm.

(C) $F_E = k\dfrac{q_1 q_2}{r^2}$. For the forces to cancel, they must be equal and opposite. Since k and q_1 are the same when calculating the force, the ratio of $\dfrac{q_2}{r^2}$ must be the same:

$$\dfrac{2\ \mu C}{(4\ cm)^2} = \dfrac{q_2}{(8\ cm)^2}, \text{ so } q_2 = 8\ \mu C.$$

Questions 138–142

138. Due to the symmetry of the arrangement of charges, the force on $-q$ will be to the left along the x-axis. Only the x-component of the force needs to be calculated: $F_E = k\dfrac{q_1 q_2}{r^2}$.

The radius between the charges is $r^2 = d^2 + d^2 = 2d^2$.

Therefore, the force between $+q$ and $-q$ is $F_E = k\dfrac{q^2}{2d^2}$.

The x-component of this force is $F_{Ex} = F_E \cos\theta$,

where $\cos\theta$ is equal to $\cos\theta = \dfrac{\text{adjacent side}}{\text{hypotenuse}} = \dfrac{d}{\sqrt{2d^2}} = \dfrac{1}{\sqrt{2}}$.

Therefore, $F_{Ex} = F_E \cos\theta = \dfrac{F_E}{\sqrt{2}}$.

There are two charges with this force acting on $-q$. So the net force on $-q$ is $\sum F_{Ex} = 2\left(\dfrac{F_E}{\sqrt{2}}\right) = \sqrt{2}F_E = \dfrac{\sqrt{2}k(q^2)}{(2d^2)}$.

This net force is in the negative x-direction, as shown in the figure.

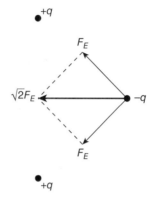

139. The net force is zero because the two electric forces are equal in size and opposite in direction.

140. As $-q$ is moved out toward $+2d$, the force from both the $+q$ charges begins to have a larger x-component. This tends to make the net force bigger on $-q$. However, as $-q$ is moved out toward $+2d$, the magnitude of the forces from both the $+q$ charges begins to decrease in magnitude because $F_E \propto \dfrac{1}{r^2}$. This tends to make the net force smaller on $-q$.

141. The charge $-q$ will accelerate to the left along the x-axis until it reaches the origin, where the net force is zero. The charge will continue moving to the left past the origin but will begin to slow down until it stops at $-2d$. The charge will oscillate along the x-axis between $+2d$ and $-2d$.

142. The charge $+q$ will accelerate to the right along the $+x$-axis and will reach a final maximum speed very far away from the origin.

Forces: The y-components of the electric forces will cancel out due to symmetry. The x-component of the electric force decreases with increasing radius $\left(F_E \propto \dfrac{1}{r^2}\right)$. Therefore, the charge $+q$ will eventually stop accelerating when it is far away from the origin and reach a maximum speed in the $+x$-direction.

Energy: The system that includes all three charges has an initial electric potential energy. When the charge $+q$ is released, the initial electric potential energy of the system converts into kinetic energy for the released charge. This means the released charge will have a maximum velocity dependent on the initial stored electric potential energy of the system ($U_{E\,initial} = K_{final\ of\ the\ released\ charge}$).

143. **(A)** $F_B > F_C > F_A = 0$.

(B) $F_E = k\frac{q_1 q_2}{r^2}$. The radius between the charges Q and point C are $r = \frac{d}{\sqrt{2}}$. Therefore, the magnitude of the force between the charges Q and the electron is $F_E = k\dfrac{Qe}{\left(\dfrac{d}{\sqrt{2}}\right)^2} = k\dfrac{2Qe}{d^2}$.

The forces from the upper right and lower left charges cancel out. The forces from the charges in the upper left and the lower right add together because they are in the same direction. Thus, the net force is $F_E = 2\left(k\dfrac{2Qe}{d^2}\right) = k\dfrac{4Qe}{d^2}$.

The direction of the force is directly away from the negative charge $-Q$.

(C) $a = \dfrac{\sum F}{m} = \dfrac{k\dfrac{4Qe}{d^2}}{m_e} = k\dfrac{4Qe}{m_e d^2}$.

144. **(A)** Note: To get full credit, all the force vectors should be the same length. Two are diagonal toward the $+q$ charges, and one is directly down and away from the $-q$ charge. The net electric force is downward.

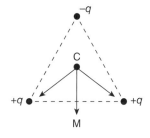

(B) The force on the proton at M is less than the force on the electron at C. At point M, the forces from the two positive charges $+q$ are equal and opposite and cancel out. In addition, the force from the negative charge $-q$ is smaller in magnitude at point M than at point C. The charges of an electron and proton are the same magnitude and have no influence on the answer.

145. **(A)** $E_A > E_B = E_D > E_C$. Electric potential energy is less negative the farther away the balloon gets from the sphere. Compare this situation to objects being attracted together by gravity. Realize that if the balloon is released, it will be attracted by the positive sphere and accelerate toward that sphere. When placed at location A, the balloon will acquire the most kinetic energy by the time it reaches the sphere. Therefore, the balloon-sphere system must have had more stored electric potential energy when at location A (the farthest point away from the metal sphere).

(B) $E_C > E_B = E_D > E_A$. Electric potential energy is the most positive when it is closest to the metal sphere. Compare this situation to compressing a spring. The more the spring is compressed, the more potential energy is stored in the system. The positive balloon is forced away from the positive sphere. The closer the balloon is to the sphere, the higher the stored energy in the balloon-sphere system.

146. Similarities: Both fields are $1/r^2$ relationships. Both fields are directed radially. Differences: Gravitational fields only point inward toward the mass creating the field. Electric fields point both toward and away from the charges, depending on the sign of the charge. The electric field strength is proportionally much stronger than gravitational fields due to the much larger Coulomb's law constant compared to the universal gravitational constant.

Questions 147–150

147. The electric fields at points N and P must be the same size. The electric field at O must be longer than the other two. See figure.

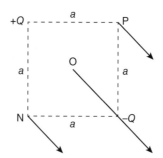

148. Remember that electric potential is a scalar without direction. Simply sum up the individual potentials: $V_P = k\dfrac{(+Q)}{a} + k\dfrac{(-Q)}{a} = 0$. The electric potential will be zero at all three points because the charges are opposite in sign, and the radius is the same in each case. All three points are on the zero equipotential line for the two-charge system.

149. Work equals the negative change in electric potential energy. Since there is no change in electric potential moving from point P to point O, there is no change in the potential energy of the proton. Therefore, the work equals zero. In addition, the charge is moved perpendicular to the electric field; therefore, no work is done by the field.

Answers < 271

150. One way to accomplish this is to move −Q to point P. The electric fields from the two charges will combine to create a net E-field to the right. See figure.

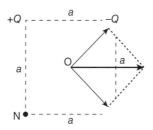

151. Here are two possible examples. Note the symmetry of the isolines and the electric potential values since the magnitude of the charges in each case is the same. Also, note how the magnitudes of electric potential decrease with distance from the charges. Electric potential will reach a value of zero at a radius of infinity (very far away).

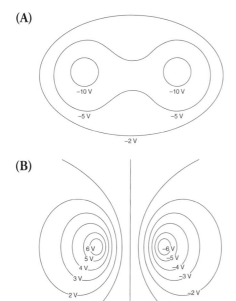

152. **(A)** Charges 1 and 3 are positive. Charge 2 is negative. The electric field vectors point toward negative charges and away from positive charges.

(B) The force will be in the opposite direction of the electric field.

(C) The isolines should be perpendicular to the electric field vectors.

(D) The isoline through point A will be at a higher electric potential because it is closer to the positive charge. In addition, electric field lines point toward lower electric potential.

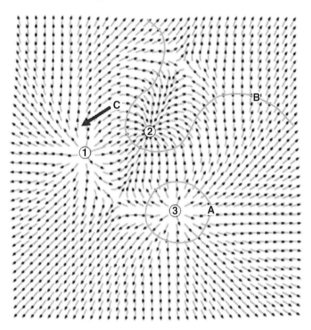

Questions 153–159

153. The isolines closest to 1 are positive; therefore, 1 is positive. The isolines nearest 2 are more negative, so 2 is negative.

154. No! There is only one kind of mass—positive. There are two types of charge—positive and negative. Since this figure has both types of charge, the isolines associated with it will not be similar to gravitational potential isolines.

155. (A) Positive charges will naturally move toward more negative electric potential areas because the electric field will point in that direction. Therefore, the proton will end up 40 V lower in potential than it started; this will be the −20 V isoline.

(B) i. In the system that includes the two charges and the proton, the electric potential energy stored in the system decreases and converts to kinetic energy for the proton.

ii. In the system that includes only the proton, the external electric field from the two charges does positive work on the proton, giving it kinetic energy.

156. Decreased. $\Delta U_E = q\Delta V$. There is a positive change in electric potential, but the charge is negative. This gives us a negative change in electric potential energy.

157. $E_{average} = \dfrac{\Delta V}{\Delta r} = \dfrac{\Delta V}{d}$. The electric field points toward decreasing electric potential, which is to the right.

158. $W = -\Delta U_E = -q\Delta V = -Q\Delta V$, or

$$W = Fd = Eqd = -\dfrac{\Delta V}{d}qd = -Q\Delta V.$$

159.

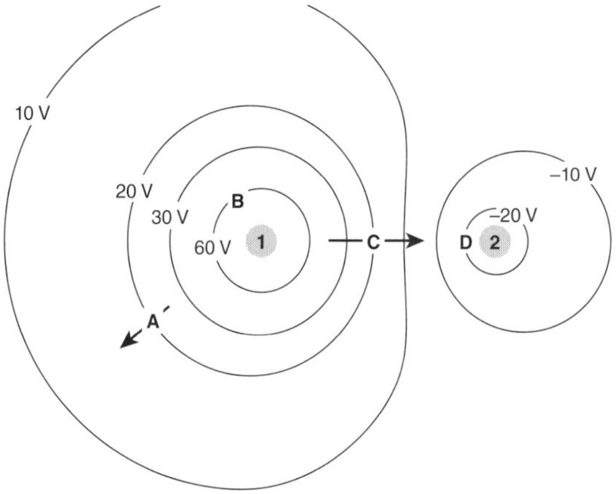

[Figure: Concentric equipotential lines around charge 1 (labeled 60 V, 30 V, 20 V, 10 V with point B and arrow A) and around charge 2 (labeled −20 V, −10 V with point D), with arrow C pointing from 1 to 2.]

The electric field vectors are always perpendicular to the isolines and point from more positive to less positive electric potential. The greater the change in electric potential per distance, the greater the electric field: $E_{average} = \dfrac{\Delta V}{\Delta r}$. Therefore, the arrow at point C should be longer than the one at point A.

160. (A) The electric force on both charges is the same in magnitude but opposite in direction (Newton's third law). The electric forces point outward.

(B) The accelerations of the two particles are not the same. Particle B has a greater acceleration due to its smaller mass.

(C) Conservation of energy for the two-particle system (note that electric potential energy is zero when the particles are very far away from each other):

$$E_{inital} = E_{final}$$

$$U_{E\ initial} = (K_A + K_B)_{final}$$

$$k\dfrac{Q(4Q)}{x} = \dfrac{1}{2}m_A v_A^2 + \dfrac{1}{2}m_B v_B^2$$

Conservation of momentum for the two-particle system (note that the particles begin with no initial momentum, so the final velocities must be in opposite directions):

$$p_{initial} = p_{final}$$
$$0 = m_A v_B + m_A v_B$$

Questions 161–165

161. The electric field inside the sphere is zero. Outside the sphere, the electric field is proportional to $1/r^2$; therefore, the graph should pass through the point $(2R, \frac{1}{4}E_0)$.

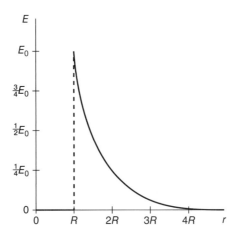

162. The electric potential inside the sphere is constant at the same value as on the surface. Outside the sphere, the electric potential is proportional to $1/r$; therefore, the graph should pass through the points $(2R, \frac{1}{2}V_0)$ and $(4R, \frac{1}{4}V_0)$.

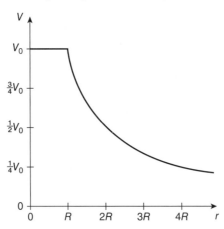

163. The electric potential of the two spheres will be the same. Charge flows until the electric potential is the same everywhere on the surface and inside the connected conductors.

164. Sphere 2 will have a larger charge than sphere 1. Sphere 2 has more surface area for the charge to spread out on. The electric potentials for the spheres must be the same at equilibrium:

$$V_1 = V_2$$
$$k\frac{Q_1}{r_1} = k\frac{Q_2}{r_2}$$
$$\frac{Q_1}{R} = \frac{Q_2}{2R}$$
$$2RQ_1 = RQ_2$$
$$2Q_1 = Q_2$$

Therefore, Q_2 is twice as big as Q_1. Applying conservation of charge, the final charges of the two spheres must equal the original charge of the system:

$$Q_{initial} = Q_{final}$$
$$Q_0 = Q_1 + Q_2$$

With these two equations, we can determine the final charge of both spheres.

165. The electric field of the smaller sphere is larger than that of the larger sphere. The charge is shared between the spheres so the electric potential is the same. This is a $1/r$ relationship. Therefore, Q_2 is twice as big as Q_1:

$$\frac{Q_1}{R} = \frac{Q_2}{2R}$$
$$2Q_1 = Q_2$$

Electric field is a $1/r^2$ relationship, so the electric field for sphere 2 is smaller than that for sphere 1:

$$E_1 = k\frac{Q_1}{R^2} > k\frac{Q_2}{(2R)^2} = k\frac{2Q_1}{4R^2} = E_2$$

The electric field of sphere 2 is half that of sphere 1.

166. (A)

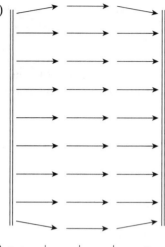

(B)

[vertical dashed equipotential lines between plates]

(C) $E = \dfrac{\Delta V}{\Delta r} = \dfrac{\Delta V}{d}$

(D) $Q = C\Delta V = \left(\varepsilon_0 \dfrac{A}{d}\right)\Delta V$

(E) The net charge is zero because the charges on the two plates are the same magnitude but opposite in sign.

(F) $F_E = Eq = \left(\dfrac{\Delta V}{d}\right)e$. The gravitational force is usually much smaller than the electric force and can be ignored in this case, because the magnitude of the proton mass is much smaller than the magnitude of the charge. The only time we need to worry about gravity is when the magnitude of the mass is much larger than the net charge of the object.

(G) Using forces:

$$a = \frac{F_E}{m_p} = \frac{Eq}{m_p} = \frac{\Delta Ve}{m_p d}$$

$$v^2 = v_0^2 + 2ad = \frac{2\Delta Ve}{m_p}$$

$$v = \sqrt{\frac{2\Delta Ve}{m_p}}$$

Using energy:

$$\Delta K = \Delta U_E$$

$$\frac{1}{2} m_p v^2 = \Delta Vq = \Delta Ve$$

$$v = \sqrt{\frac{2\Delta Ve}{m_p}}$$

The answers for forces and energy are the same.

(H) The proton released from the center of the capacitor has a smaller final velocity than that of the proton released from the negative plate. Both have the same acceleration: $a = \frac{F_E}{m_p} = \frac{Eq}{m_p} = \frac{\Delta Ve}{m_p d}$. But the one released from the middle of the capacitor has less distance to accelerate: $v^2 = v_0^2 + 2ad$.

Or we can just say that the proton released from the middle of the capacitor moves through a smaller electric potential, thus reducing its final velocity.

167. (A) Moving the electron toward the negative plate will increase its potential energy. Paths 3, 5, and 8 all produce the same increase in electric potential energy because they all move through the same change in electric potential.

(B) We need the largest change in electric potential to produce the greatest final velocity. Since the proton will move to the right, we need to release the proton from any of the points 1, 4, or 6.

(C) The electric field inside a capacitor, away from the edges, is uniform and constant. All the points have the same electric field strength.

Questions 168 and 169

168. Both will travel through a parabolic trajectory. The baseball will arc downward toward the Earth, while the proton will arc to the left toward the negative plate. The acceleration of the baseball will be g. The acceleration of the proton will be $a = \dfrac{F_E}{m_p} = \dfrac{Ee}{m_p}$.

169. The electron has a negative charge and will arc toward the positive plate on the right. The acceleration of the electron will be greater than that of the proton due to its smaller mass. This means it will curve faster in a tighter parabolic path toward the right.

Questions 170–174

170. The two alterations cancel each other out, and the capacitance will remain the same: $C = \kappa\varepsilon_0 \dfrac{A}{d} = \kappa\varepsilon_0 \dfrac{2A}{2d}$.

171. The capacitance remains the same as it is determined by the geometry of the capacitor itself: $C = \kappa\varepsilon_0 \dfrac{A}{d}$. Since the capacitance stays the same, the charge on the plates will double: $2Q = C(2\Delta V)$.

172. Capacitance is directly proportional to the plate area, $C = \kappa\varepsilon_0 \dfrac{A}{d}$, so the graph is a line with a positive slope.

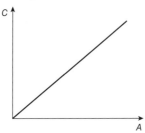

173. Capacitance is inversely proportional to the distance between the plates, $C = \kappa\varepsilon_0 \dfrac{A}{d}$. Therefore, the graph is a hyperbola.

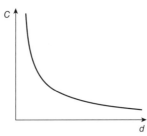

174. We need the most stored energy to light the bulb for the longest time: $U_C = \frac{1}{2}C(\Delta V)^2$. We need the largest capacitance possible: $C = \kappa\varepsilon_0 \frac{A}{d}$. Therefore, we need the largest dielectric constant, the largest plate area, and the smallest plate spacing we can get.

Questions 175 and 176

175. The dipole will turn clockwise due to a clockwise torque. However, the dipole would not propagate to the right or left in the electric field because the force on the positive and negative charges are equal in magnitude.

176. (A) The box will polarize with a net negative charge on the right and a net positive charge on the left. Note that the total charge of the box remains zero due to conservation of charge.

(B) The electric field due to the polarization of the box will be to the right.

(C) The net electric field inside the box will be zero. This is an example of electromagnetic shielding.

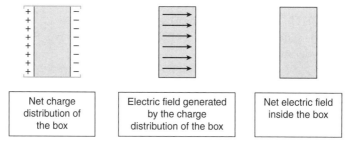

AP-Style Multiple-Choice Questions

177. (C) Only electrons can move in the metal spheres. Protons (positive charges) are stuck in the nuclei of atoms and cannot move. In the presence of the positively charged rod, the three-sphere system will polarize with excess electrons moving toward the positively charged rod. Answer choice A shows no charge polarization. Choice B shows each individual sphere polarized, which cannot happen since the conductors are in contact and act as one system. Choice D shows protons moving, which also cannot happen.

178. (D) Electric field lines are perpendicular to the isolines and point from higher potential to lower potential. The average electric field strength is $E_{average} = \frac{\Delta V}{\Delta r}$. Therefore, the electric field at point A will be stronger than at point B.

179. (A) Gravity is proportional to the product of the two masses, and the electric force is proportional to the product of the two charges. Since the mass of an electron is on the order of 10^{-31} kg, and the charge is on the order of 10^{-19} C, the electric force will be much larger than the gravitational force. In addition, the universal gravitational constant is much smaller than the Coulomb's law constant. This makes the gravitational force between the electrons negligible compared to the electric force.

180. (D) Without knowing the exact locations, it is impossible to know the exact electric field strength. Since electric field is proportional to the inverse of the radius squared, the electric field strength varies in strength in the three regions.

181. (D) The charged balloon will polarize both the wooden board and the steel plate. Therefore, it will be attracted to both. However, the polarization of the wood occurs on an atomic scale because it is an insulator, and its electrons do not move easily. The steel is a conductor that allows its electrons to migrate. This permits the electrons in the steel to move farther and create a larger charge separation in the process of polarization. This means the balloon will be attracted to the steel more strongly than to the wood.

182. (D) The rod will push negative charges to the far side of the sphere. The side of the sphere close to the rod will become more positive. This process is called charge polarization.

183. (B) The charge of Psevdísium is smaller than the electron charge, which calls this particle's existence into doubt. Within the level of uncertainly listed, the charge of Alithísium is ten times the charge of the electron. We would expect the charge to be a whole integer multiple of the electron charge. The mass of Alithísium is listed as an energy equivalent. This is perfectly acceptable: $E = mc^2$.

184. (A and C) Sphere A is negative, and sphere B is positive. The rod polarizes the system that consists of the two spheres, pulling excess electrons to the left and leaving the right side with an excess positive charge of equal magnitude.

185. (C) The force from sphere 1 on sphere 3 is ½F to the left. The force from sphere 2 on sphere 3 is 2F to the left. The sum is (5/2)F.

186. (C) For the electron to receive a force to the left, the electric field must be pointing to the right. The electric field from the charge at vertex B must be smaller in magnitude than the field produced by the charge at vertex A. To accomplish this, the charge at point B must be negative and smaller than $|+q|$.

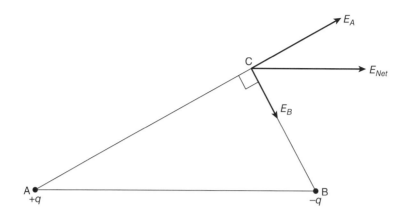

187. (A) $\Delta U_E = q\Delta V$. To get the greatest increase in electric potential energy, we need the greatest change in electric potential times the charge. The charge of protons and electrons are the same magnitude. To increase the electric potential energy of a proton, we need to move the proton to higher potentials. To increase the electric potential energy of the electron, we need to move the electron to lower electric potentials.

188. (C) Electric field vectors are always perpendicular to the equipotential lines. The pattern of the electric field vectors indicates that both charges are the same sign. Additionally, there is no zero potential line separating the two charges, indicating that they have the same sign. The equipotential lines indicate that the charge on the right has a larger magnitude.

Questions 189–193

189. (B) The electric field between the plates of a parallel plate capacitor is uniform and constant in strength as long as you are not too close to the edges of the capacitor.

190. (A) Near the edges of the capacitor, the electric field is not uniform and is weaker. Thus the electric force on e_1 is smaller in magnitude than on the other charges. The electric field forces both electrons to the right and the protons to the left.

191. (A) The electric force on a positive charge is in the direction of the electric field. The gravity force is much smaller than the electric force. All the other trajectories show gravity to be a force similar or larger in magnitude to the electric force.

192. (B) Both e_2 and p_2 will travel through the same distance of $3x$, which is also the largest potential difference. Both also receive the same magnitude of electric force. The mass of an electron is much smaller than that of a proton. Therefore, the electron will achieve a greater final velocity.

193. (B and C)

$$E = \frac{\Delta V}{\Delta r} = \frac{V_{final} - V_{initial}}{3x} = \frac{V_0 - V_{3x}}{3x}$$

$$E = \frac{\Delta V}{\Delta r} = \left(\frac{1}{\Delta r}\right)\left(\frac{\Delta U}{q}\right) = \left(\frac{1}{3x}\right)\left(\frac{\Delta K}{e}\right) = \left(\frac{1}{3x}\right)\left(\frac{\frac{1}{2}m_p v^2}{e}\right) = \frac{m_p v^2}{6xe}$$

AP-Style Free-Response Questions

194. (a) The net electric field is the vector addition of the two separate electric fields created by $-2q$ and $+q$.

$$E_{-2q} = k\frac{2q}{x^2} \text{ to the right}$$

$$E_{+q} = k\frac{q}{(3x)^2} = k\frac{q}{9x^2} \text{ to the left}$$

$$E_{Total} = k\frac{2q}{x^2} - k\frac{q}{9x^2} = k\frac{q}{x^2}\left(2 - \frac{1}{9}\right) = \frac{17kq}{9x^2} \text{ to the right}$$

(b) The electric force on a charge in an electric field is $F_E = Eq$. Positive charges receive a force in the direction of the electric field. Negative charges receive a force in the opposite direction of the field.

$$F_E = Eq$$

$$F_E = \left(\frac{17kq}{9x^2}\right)(5q) = \frac{85kq^2}{9x^2} \text{ to the right}$$

(c) $E \propto \frac{q}{x^2}$. Since the $-2q$ is twice the charge and three times the distance of the $+q$ charge, the electric field will be dominated by the $+q$ charge and the electric field will be to the right.

(d) $E_0 > E_{-2x} > E_{+2x}$

(e) The electric field is directed to the right beyond $+x$ and $-x$. Between the two charges, the electric field is to the left. Note that the field is asymptotic at a vertical line through both $+x$ and $-x$.

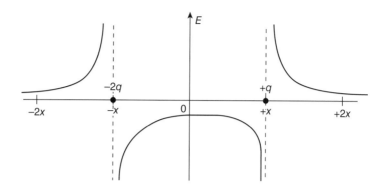

(f) The electric fields from the two charges are in the opposite directions beyond +x and −x. Since the left charge is larger than the right charge, there will only be a zero electric field location to the right of the +q charge. This means that our graph in part (d) will eventually cross the horizontal axis and become negative somewhere to the right of point +q.

195. (a) The electric field is stronger at point M. Due to symmetry, the field cancels out to zero at C. At point M, the electric fields from the bottom two charges cancel out, leaving a downward field from the top charge.

(b) The electric field is downward at point M. If an electron is placed there, it will receive an electric force upward.

(c) i. $\Delta U_E = Q\Delta V = Q(V_{final} - V_{initial}) = Q(V_C - V_M)$

ii. $W = \Delta U_E = Q\Delta V = Q(E_{average}\Delta r) = QE_{average}x$

(d) No.

   i. The electric field decreases with the square of the distance. Therefore, the electric force on the released charge from the other two charges will eventually decrease to zero when very far away. At that point, the charge will no longer be accelerating. The released charge reaches a maximum speed when it is far enough away from the other two charges not to feel their repulsive influence anymore.

   ii. By conservation of energy, the electric potential energy stored in the system of three charges converts to kinetic energy of the released charge. Since the stored electric potential energy of the system is not infinite, the released charge cannot accelerate forever to an infinite kinetic energy. When very far away from the other two charges, the released charge will have reached its maximum speed.

196. **(a)** The electric field is to the right. The electron is accelerating to the left. Therefore, it is receiving an electric force to the left. The electric field exerts a force in the opposite direction on negative charges:
$$E = \frac{\Delta V}{\Delta r} = \frac{\Delta V}{d} = 25{,}000 \; \frac{V}{m}.$$

(b) $K_{initial} = \Delta U$

$$\frac{1}{2} m_e v_{initial}^2 = e \Delta V$$

$$v_{initial} = 1.33 \times 10^7 \; \frac{m}{s}$$

(c) $a = \dfrac{\sum F}{m} = \dfrac{F_E}{m_e} = \dfrac{Eq}{m_e} = \dfrac{Ee}{m_e} = 4.39 \times 10^{15} \; \dfrac{m}{s^2}$

$v = v_0 + at$

$t = \dfrac{\Delta v}{a} = 3.03 \times 10^{-9} \, s = 3.03 \; ns$

(d) $E = \dfrac{Q}{\varepsilon_0 A}$

$Q = \varepsilon_0 A E = 6.64 \times 10^{-8} \, C$

(e) The path should be parabolic.

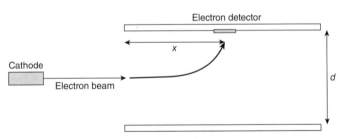

(f) The *x*-velocity should be constant. The *y*-velocity should have a positive slope starting at zero.

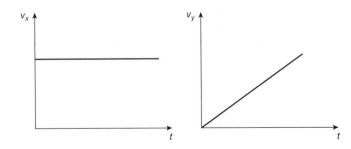

(g) Since the alpha particle is positive, it will curve in the opposite direction of the electron and strike the bottom plate. The alpha particle has two times the charge of an electron. Since the electric field is the same, the electric force on the alpha particle will be two times that of the electron. This will tend to increase the acceleration of the alpha particle toward the bottom plate. However, the alpha particle is much heavier than the electron. This will outweigh the effect of the greater electric force, causing the acceleration of the alpha particle to be less than that of the electron. Therefore, the detector should be placed on the bottom plate at a distance greater than x to locate where the alpha particles strike the plate.

Chapter 4: Electric Circuits

Skill-Building Questions

197. Electric current is the physical movement of charge carriers. In a solid material, the charge carriers are electrons that move through the solid in the opposite direction of the electric field. This means the movement is from lower, more negative potentials toward higher, more positive potentials. Conventional current assumes positive charge carriers that move in the direction of the electric field from high to low electric potentials.

198. Electric potential is the electric potential energy per charge: $V = U/q$. Electric potential difference is literally the difference in electric potential between two locations: $\Delta V = (V_2 - V_1)$. This is also commonly referred to as voltage. emf is the potential difference provided by a battery, solar cell, generator, or other power source.

Questions 199–202

199.

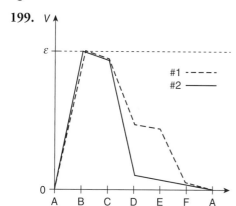

200. The slanted region between C and D would be longer and descend to a lower final point because the potential consumed in the resistor would be greater. This would leave less potential for the resistor between points E and F.

201. Wires with negligible resistance would not consume any potential. The lines between points B and C, D and E, and F and A would be flat, with a slope of zero.

202. See figure for problem 199 above. From point A to point C, the graphs are the same. With only one resistor, the entire emf of the battery is consumed between points C and D. This makes the line steeper than the dashed graph between C and D.

Questions 203–206

203. Note that the current is constant because there is only one pathway.

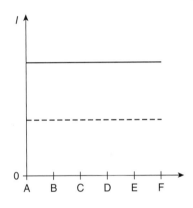

204. This would increase the total resistance of the circuit from 2R to 3R. This would bring the dashed line to a lower constant value of two-thirds the original current.

205. With wires of negligible resistance, the total resistance of the circuit would decrease a little, causing the dashed line to be at a slightly higher constant value.

206. See figure for problem 203 above. Removing one resistor cuts the resistance in half, which doubles the current.

Questions 207–212

207.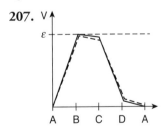

208. There would be no change because the resistor has to consume all of the potential between points C and D.

209. See figure for problem 207 above. The graph for path #2 is the same as for path #1. Both paths consume the same potential.

210. The current is not constant because it splits between the two paths.

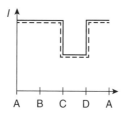

211. Twice the resistance would cut the current through path #1 to half its original value. So the dashed line C-D would be half as high. This would also affect the current from A-B-C and D-A because this is the wire that feeds path #1. Therefore, all parts of the graph would move downward by the same amount.

212. See figure for problem 210 above. The graph of the current through path #2 is identical to that for path #1 because both paths are identical with the same resistance.

213.

Circuit element	Point or pair of points to which you would connect a voltmeter to measure the potential difference across the circuit element	Point or pair of points to which you would connect an ammeter to measure the current through the circuit element
Battery in circuit #1	Pairs that work: A and B, A and C, B and F, C and F	Points that work: A, B, C, D, E, F
Top resistor in circuit #1	B and D, C and D, C and E	A, B, C, D, E, F
Battery in circuit #2	A and B, B and E, B and F	A, B
Far right resistor in circuit #2	C and A, C and E, C and F, D and A, D and E, D and F	D, F

214. Ohmic materials have a constant resistance that is independent of the voltage applied or current passing through them. Non-ohmic materials have a resistance that is dependent on the voltage applied or current passing through them.

215. Calculate the resistance over a wide range of voltages and currents to see whether or not the resistance is constant. An easy way to do this is with a graph of current as a function of voltage. Simply see if the graph maintains a constant slope.

216. **(A)** $R = \dfrac{\rho L}{A}$. The resistance doubles.

(B) The resistance is one-third of the original.

(C) Area is proportional to the radius squared. The resistance is one-sixteenth of the original.

217. The loop rule is an application of conservation of energy. The energy (voltage) supplied in a loop must be completely consumed in the loop. The electric potential differences across the components in a circuit loop must always add up to zero: $\sum \Delta V_{loop} = 0$.

218. The junction rule is an application of conservation of charge. Charge is not created or destroyed in a circuit. The same amount of charge that enters a junction (positive current) must exit that junction (negative current). Therefore, the net current at any junction must sum to zero: $\sum I_{junction} = 0$.

219. **(A)** Since there is only one pathway, Kirchhoff's junction rule shows us that the current must be the same in each resistor.

(B) $\varepsilon - I_1 R_1 - I_1 R_2 - I_1 R_3 - I_1 R_4 = 0$

(C) $R_{eq} = R_1 + R_2 + R_3 + R_4$

(D) $I = \dfrac{\Delta V}{R_{eq}} = \dfrac{\varepsilon}{R_1 + R_2 + R_3 + R_4}$

(E) $\Delta V_2 = I R_2 = \dfrac{\varepsilon R_2}{R_1 + R_2 + R_3 + R_4}$

220. **(A)** Using Kirchhoff's loop rule to write the equation for the loop that contains the battery and R_2, we get $\varepsilon - I_2 R_2 = 0$. Writing the loop rule equation for the loop that contains the battery and R_1, we get $\varepsilon - I_1 R_1 = 0$. Comparing the two equations, we see that the potential difference across both resistors must be equal to the emf of the battery.

(B) $\varepsilon - I_1 R_1 = 0$ and $\varepsilon - I_2 R_2 = 0$

(C) For the junction at the top of the circuit: $I_3 - I_1 - I_2 = 0$

(D) $R_{eq} = \dfrac{R_1 R_2}{R_1 + R_2}$

(E) $I_1 = \dfrac{\Delta V}{R_1} = \dfrac{\varepsilon}{R_1}, \; I_2 = \dfrac{\Delta V}{R_2} = \dfrac{\varepsilon}{R_2}, \; I_3 = \dfrac{\Delta V}{R_{eq}} = \dfrac{\varepsilon}{R_1} + \dfrac{\varepsilon}{R_2} = \dfrac{\varepsilon(R_1 + R_2)}{R_1 R_2}$

221. **(A)** For the junction of the left: $I_1 - I_2 - I_3 = 0$. For the junction of the right: $I_3 + I_2 - I_4 = 0$; therefore, $I_1 = I_4$.

(B) $\varepsilon - I_1 R_1 - I_2 R_2 - I_2 R_3 = 0$

(C) $I_1 = I_4 > I_3 > I_2$. Currents 1 and 4 are the same and equal to the combination of currents 2 and 3. Currents 2 and 3 have the same potential but different resistances. The resistance is less for current 3; therefore, current 3 is larger than current 2.

(D) $R_{eq} = R_1 + \dfrac{(R_2 + R_3) R_4}{R_2 + R_3 + R_4}$. Since all the resistors are identical, we can simplify this expression to be $R_{eq} = 5R/3$.

222. (A) $P = \dfrac{V^2}{R}$, $R_{40W} = 360\ \Omega$, $R_{100W} = 144\ \Omega$.

(B) $P = \dfrac{V^2}{R}$. When connected in parallel, the bulbs receive the same electric potential. Therefore, the bulb with the smallest resistance (100 W bulb) is brighter.

(C) $P = I^2 R$. When connected in series, the bulbs receive the same current. Therefore, the bulb with the largest resistance (40 W bulb) is brightest.

Questions 223–226

223. $\varepsilon - Ir - IR = 0$

224. $r = \dfrac{\varepsilon - IR}{I}$. Since we are solving for two unknowns with only one equation, we would need to know two sets of current I and the resistance R paired numbers: (I_1, R_1) and (I_2, R_2). We would need to measure the current through the system for two different external resistances R.

225. $V_{AB} = V_{terminal} = \varepsilon - Ir$. This equation shows us that when the current through the battery is zero, the potential difference at the terminals of the battery equals the emf of the battery.

226. Examination of the terminal voltage equation shows us that $V_{terminal}$ is a linear function of the current I. The slope of the line will equal $-r$. The intercepts represent the emf of the battery when the current is zero and the maximum current the battery can put out when the external resistance is zero. See figure.

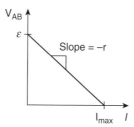

227. C > A = B, $P = \dfrac{V^2}{R}$. All the bulbs have the same resistance. Therefore, the bulb that receives the greatest potential difference will have the greatest power consumption and be brightest.

228. D > A > C = B, $P = I^2 R$. All the bulbs have the same resistance. Therefore, the bulb that receives the greatest current will have the greatest power consumption and be brightest. Bulb D is in the main current pathway and receives the greatest current. The main current splits between the two pathways. Bulb A sits in the pathway with the least resistance, so it will receive more current than bulbs C and B. Bulbs C and B are in the same pathway and receive the smallest current.

229. $P = IV$. Power is a direct function of current. Therefore, the slope of the graph is the electric potential: $V = 12$ V.

230. $P = I^2 R$. Power is directly related to the current squared. Therefore, the slope of the graph is the resistance. $R = 50\ \Omega$.

231. $P = \dfrac{V^2}{R}$. Power is directly related to the electric potential squared. Therefore, the slope of the graph is equal to $\dfrac{1}{R}$. $R = 20\ \Omega$.

232. There is more than one way to draw each of these.

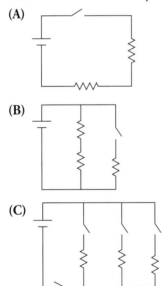

Answers 291

233. There is more than one way to draw each of these.

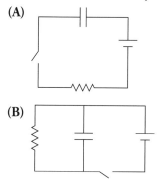

234.

Component	V	I	R	P
R_1	6 V	1.5 A	4 Ω (on the right)	9 W
R_2	6 V	1.0	6 Ω	6 W
R_3	2 V	0.5 A	4 Ω (on the left)	1 W
R_4	4 V	0.5 A	8 Ω	2 W
Total for circuit	12 V	1.5 A	8 Ω	18 W

235. Note that numbers in the table are rounded to two significant digits.

Component	V	I	R	P
R_1	82 V	2.7 A	30 Ω	220 W
R_2	18 V	0.91 A	20 Ω	17 W
R_3	18 V	1.8 A	10 Ω	33 W
Total for circuit	100 V	2.7 A	37 Ω	270 W

236. Note that numbers in the table are rounded to two significant digits.

Component	V	I	R	P
R_1	1.3 V	0.22 A	6 Ω	0.29 W
R_2	1.3 V	0.43 A	3 Ω	0.59 W
R_3	8.0 V	0.67 A	12 Ω	5.4 W
R_4	11 V	0.67 A	16 Ω	7.2 W
Total for circuit	20 V	0.67 A	30 Ω	13 W

237. The total resistance of the circuit decreases when the switch is closed. The reading on ammeter 1 increases because it measures the total current. Ammeter 2 receives half of the new current through ammeter 1. Since the total resistance has not been cut in half when the switch is closed, the total current has not doubled. Therefore, the reading of ammeter 2 must go down.

238. (A) The capacitances add up to 6 µF.

(B) $U = \frac{1}{2}C(\Delta V)^2 = 0.08$ J

239. (A) 6 µF (Remember that the rules for adding capacitors is the opposite of those for resistors!)

(B) $U = \frac{1}{2}C(\Delta V)^2 = \frac{1}{2}(2 \text{ µF})(200 \text{ V})^2 = 0.04$ J

240. When the switch is first closed, the capacitor behaves like a wire. This creates a short circuit around ammeter 1, and it will read zero. All the current is flowing through only two resistors in series and ammeter 2. When the capacitor becomes fully charged, it behaves like an open switch in the circuit. The current is now passing through all three resistors in series. This increases the current in ammeter 1 but decreases the current in ammeter 2 because the total resistance of the circuit has increased.

241. (A) Immediately after the switch is closed, the capacitors acts like a short circuit wire.

Location	V	I	R	P
1	0	0	15 Ω	0
2	0	0	10 Ω	0
3	12 V	6 A	2.0 Ω	72 W
Total for circuit	12 V	6 A	2.0 Ω	72 W

(B) After a long time the capacitor is "full" and behaves like an open switch in the circuit.

Location	V	I	R	P
1	9.0 V	0.60 A	15 Ω	5.4 W
2	9.0 V	0.90 A	10 Ω	8.1 W
3	3.0 V	1.5 A	2.0 Ω	4.5 W
Total for circuit	12 V	1.5 A	8.0 Ω	18 W

242. **(A)** Immediately after the switch is closed, the capacitors acts like a short circuit wire.

Location	V	I	R	P
1	12 V	0.80 A	15 Ω	9.6 W
2	12 V	1.2 A	10 Ω	14 W
Total for circuit	12 V	2.0 A	6.0 Ω	24 W

(B) After a long time the capacitor is "full" and behaves like an open switch in the circuit.

Location	V	I	R	P
1	12 V	0.80 A	15 Ω	9.6 W
2	0	0	10 Ω	0
Total for circuit	12 V	0.80 A	15 Ω	9.6 W

(C) $U = \frac{1}{2}C(\Delta V)^2 = \frac{1}{2}(240 \text{ μF})(12 \text{ V})^2 = 0.017 \text{ J}$

AP-Style Multiple-Choice Questions

Questions 243 and 244

243. **(A)** There is only one pathway. The current is the same.

244. **(B)** The current is the same; therefore, the higher the resistance, the greater the electric potential. Use the resistance equation to find the relative magnitude of the three resistors: $R = \frac{\rho L}{A}$.

Questions 245 and 246

245. **(B)** Ohmic materials will have a constant ratio of current to potential difference. This shows up on a graph as a straight line.

246. **(D)** In parallel, both bulb and resistor receive the same 10 volts.

247. **(C)** $P = I\Delta V$. Calculate power for both and add.

248. **(B)** $R = \Delta V/I$. The resistance is 10 Ω for the resistor and 14.3 Ω for the bulb. Adding them in parallel gives us 5.9 Ω.

249. **(B)** This is a tricky one because the resistance of the bulb is non-ohmic, and its resistance changes with voltage and current. We know that in series the current

passing through the resistor and bulb must be the same. Looking back at our graph, we see that this occurs at a current of 0.50 A. Checking to make sure that the voltages work out, we see that the potential across each will be 5 V, which works out perfectly because in series, the voltage drops of the resistor and the bulb must add up to the voltage supplied by the battery.

Questions 250–251

250. (A and C) Applying Kirchhoff's loop rule to the bottom loop, we get answer choice A. Applying Kirchhoff's junction rule to the left junction, we get answer choice C.

251. (A) The emfs of both batteries point in the same direction for the outer loop, for a combined potential difference of 15 V across R_1. R_2 is in parallel with the ε_2. Therefore, the voltage across R_2 is only 6 V. As a consequence, less current passes through R_2.

252. (D) Originally only bulb A is lit, and it experiences all the emf of the battery. When the switch is opened, the emf of the battery is split evenly between the bulbs. $P = \dfrac{V^2}{R}$; therefore, the power dissipation by bulb A is one-quarter of the original.

Questions 253 and 254

253. (B) The three resistors in the bottom right corner are in parallel. Since they all have the same resistance and the same electric potential across them, they must also have the same current. Ammeter #3 is the sum of the currents in the top two parallel resistors and will be twice as large as ammeter #4.

254. (B) The three resistors in parallel add up to a resistance of ⅓R. Adding these in series with the resistor in the main line, we get 4/3R.

255. (D) $R = \dfrac{\rho l}{A} = \dfrac{\rho l}{\pi r^2}$. The resistor on the right has twice the radius and a quarter of the resistance of the one on the left. Therefore, the current through the right resistor will be four times larger.

256. (C and D) You may find it helpful to set up a VIR chart for this problem. Measuring ΔV between points 1 and 2 with the switch closed measures the voltage for the 100 ohm battery. Measuring the potential difference when the switch is open gives us the battery's emf. All we need is either the current or the potential difference, when the switch is closed, to find the internal resistance of the battery. With either, we can calculate the potential difference across the internal resistance and the current passing through it. Then we can find the value of the internal resistance.

257. (B) The reading in ammeter 1 never changes because the potential difference and resistance in that parallel line does not change. The capacitor behaves like a wire when uncharged and like an open switch when fully charged. This behavior changes the current through ammeter 2 as time passes.

258. (C and D) Four capacitors connected in series result in a capacitance one-quarter the size of the individual capacitors. Our equation for capacitance is: $C = \dfrac{\kappa \varepsilon_0 A}{d}$. To get a single capacitor with one-quarter the capacitance, we need to get a factor of four in the denominator.

Questions 259 and 260

259. (A and D) The batteries are connected in series; therefore, their electric potentials add up. The difference between the voltages in the left graph is 1.5 V. Thus, the batteries must be 1.5 V each. The right graph shows power proportional to the voltage squared because the graph is linear.

260. (D) $P = \dfrac{V^2}{R}$. The slope of the right graph will equal the reciprocal of the resistance.

Questions 261 and 262

261. (B) The energy supplied to the gas is related to the power of the circuit, time, and the internal energy of the gas:

$$Pt = \frac{\Delta V^2}{R} t = \frac{\varepsilon^2}{R} t = \Delta U = \frac{3}{2} N k_B \Delta T$$

$$\Delta T = \frac{2\varepsilon^2 t}{3 N k_b R}$$

262. (B) The two resistors in parallel will have half the total resistance $\dfrac{R}{2}$, which will double the current and the power output of the circuit. This will double the slope of the graph.

AP-Style Free-Response Questions

263. (a) i. Minimum equipment: wires, voltmeter, and ammeter
  ii. There are several ways to draw this. The key is that the ammeter is in series with the battery and the graphite. The voltmeter must be in parallel with the graphite to measure the correct potential difference.

The graphite should be drawn as a resistor and the entire schematic needs to be labeled.

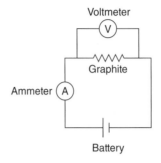

iii. Procedure:
1. Connect the graphite, ammeter, and battery in series.
2. Connect the voltmeter in parallel with the graphite.
3. Measure the current and voltage.
4. Repeat for several lengths of graphite.

iv. Since the current and voltage are measured for the graphite specifically, the details of the emf source are irrelevant.

Trial	Diameter (m)	Length (m)	Current (A)	Voltage across Play-Doh (V)	Resistance (Ω)
1	0.002	0.1	0.003	9.0	3,000
2	0.002	0.2	0.001	9.0	
3	0.002	0.3	0.001	9.0	
4	0.002	0.4	0.001	9.0	
5	0.002	0.5	0.001	9.0	
6	0.003	0.1	0.006	9.0	1,500
7	0.003	0.2	0.003	9.0	
8	0.003	0.3	0.002	9.0	
9	0.004	0.1	0.011	9.0	820
10	0.004	0.2	0.006	9.0	
11	0.006	0.1	0.025	8.9	360
12	0.008	0.1	0.045	8.9	200
13	0.010	0.1	0.069	8.8	130

(b) i. Trials: 1, 6, 9, 11, 12, and 13 are the best because they give the widest quantity and spread of data for a constant length segment of Play-Doh with differing diameters. The resistance is missing and needs to be calculated. (See the table.)

ii. The graph should be a curve, and the axes must be labeled.

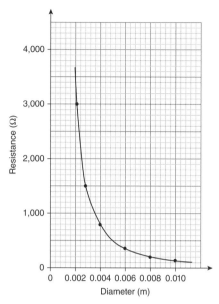

iii. It appears that resistance is inversely proportional to the diameter of the conductive path. But it is hard to tell. It could also be that resistance is inversely proportional to the diameter squared. Either answer would be appropriate based on the question and the information on display in the graph.

iv. To prove which relationship we actually have, we need to graph R – vs– 1/d and R – vs – 1/d². Whichever graph shows a straight line indicates the correct relationship.

v. The changing water content of Play-Doh could change the resistivity of the material and the resistance-diameter relationship, which damages the validity of the lab. It is a variable that is not held constant. That is a problem.

264. (a) Agree. Writing Kirchhoff's loop rule equation for the upper loop, we get

$$\varepsilon - I_2 R - I_2 R - I_1 R = 0$$
$$\varepsilon = 2I_2 R - I_1 R$$

Applying Kirchhoff's loop rule to the outer loop, we get

$$\varepsilon - I_3 R - I_1 R = 0$$
$$\varepsilon = I_3 R - I_1 R$$

Combining the equations, we have

$$2I_2R - I_1R = I_3R - I_1R$$
$$2I_2R = I_3R$$
$$2I_2 = I_3$$

(b) $P_1 > P_4 > P_2 = P_3$. Power is $P = I^2R$. All the resistors are the same. Therefore, the ranking is based on the current. Resistor 1 receives the most current as all the current must pass through it, and we just proved in answer (a) that resistor 4 receives twice the current of resistors 2 and 3.

(c) The current 1 in terms of current 3:

$$I_1 = I_2 + I_3 = \frac{1}{2}I_3 + I_3 = \frac{3}{2}I_3$$

The power of resistor 1 in terms of resistor 4:

$$P_1 = I_1^2 R = \left(\frac{3}{2}I_3\right)^2 R = \frac{9}{4}I_3^2 R = \frac{9}{4}P$$

(d) Agree that the power will go down for resistor 1: When the switch is opened, there is only one path left for the current to pass through. This means the total resistance of this new series circuit increases. The potential difference across resistor 1 will decrease, which will bring its power dissipation down as well.

Disagree that resistor 2 and 3 are unaffected: The original current passing through resistors 2 and 3 was

$$I_2 = \frac{1}{3}I_1 = \frac{1}{3}\left(\frac{\varepsilon}{R_{total}}\right) = \frac{1}{3}\left(\frac{\varepsilon}{\frac{5}{3}R}\right) = \frac{\varepsilon}{5R}$$

The new current through resistors 2 and 3 is

$$I_2 = I_1 = \frac{\varepsilon}{R_{total}} = \frac{\varepsilon}{3R}$$

The new current is larger than the old. Power dissipation goes up.

(e) i. Immediately after the switch is closed, the capacitor acts like a short circuit wire. The current through resistor 1: $I_1 = \frac{\varepsilon}{R}$.

The potential difference across the capacitor: $\Delta V = 0$.

ii. After a long period of time, the capacitor becomes fully charged and acts like an open switch in the circuit. The current through resistor 1: $I_1 = \dfrac{\varepsilon}{R_{total}} = \dfrac{\varepsilon}{3R}$.

The potential difference across the capacitor will be equal to that of resistors 2 and 3 combined, because the capacitor is in parallel with them: $\Delta V = IR = I_1(R+R) = \dfrac{\varepsilon}{R_{total}}(2R) = \dfrac{2}{3}\varepsilon$.

iii. Potential energy: $\dfrac{1}{2}CV^2 = \dfrac{1}{2}C\left(\dfrac{2}{3}\varepsilon\right)^2 = \dfrac{2}{9}C\varepsilon^2$.

265. (a) $V_T = \varepsilon - Ir$

(b) i. It is important to use data from the best fit line of the graph. All experimental data have errors. Therefore, the data in the table have experimental errors. The best fit line is the best average of the data.

The easiest way to do this is to realize that the negative of the slope of the line is the internal resistance of the battery. Taking points from the best fit line, we get an internal resistance of 1.5 Ω. With this, we can pull a point off the graph and use the equation for part (a) to solve for emf = 12 V. We can also find the internal resistance by using data from the graph to set up two simultaneous equations with unknowns of ε and r.

ii. The y-intercept of the graph is the emf of the battery.

iii. The x-intercept is the maximum current that can be produced by the battery.

(c) i. Internal resistance of the battery will not influence the energy stored in the capacitor. When the capacitor is fully charged, there will be no current flowing from the battery. At this point, the graph shows that the terminal voltage of the battery is equal to the emf. Thus, the internal resistance has no effect on the voltage across the capacitor.

ii. Capacitors in parallel add up; therefore, the combined capacitance is 2C. The energy stored will be

$$U = \dfrac{1}{2}CV^2 = \dfrac{1}{2}(2C)\varepsilon^2 = C\varepsilon^2 = 3600\ \mu J$$

iii. We need twice the capacitance of the original capacitor. We can accomplish this by doubling the plate area or cutting the distance between the plates in half: $C = \dfrac{\kappa\varepsilon_0 A}{d}$.

Chapter 5: Magnetism and Electromagnetic Induction

Skill-Building Questions

266. Similarities: There are two types of charge (positive and negative) and two poles (north and south). Opposites attract, and likes repel. Both charges and poles produce fields that influence other charges and poles. Both fields decrease with distance.

Differences: Charges can exist individually, but poles always come in pairs (north and south). Charges exert forces on all other charges, while poles only influence moving charges. Electric forces are always parallel to the field lines. Magnetic forces are perpendicular to the field lines.

267. It will produce two smaller, weaker magnets.

268. Magnetic fields are produced by moving charges. Magnetic fields always come in "loops," without beginning or end. This always produces a magnetic pair of poles.

269. The north end of a compass is attracted to the south magnetic pole of the Earth. The south magnetic pole of the Earth is in the geographic north (up where Santa lives). You can also say that the compass aligns with the Earth's magnetic field with the north end of the compass pointing in the direction of the magnetic field. The Earth's magnetic field exits the Earth in Antarctica and reenters near Canada and Russia.

270.

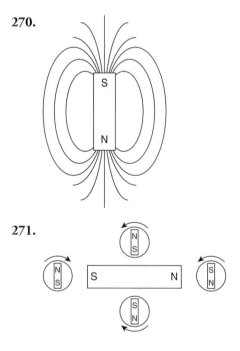

271.

272.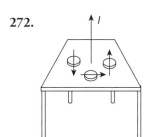

273. Both are attracted to the magnet, but not because the bar is a magnet. The positive spheres polarize the magnet and are then attracted to it. The same thing would happen if it were a block of wood or metal instead of a magnet.

274.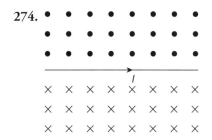

275. See figure. The arrow should be longest at A and shortest at B.

276.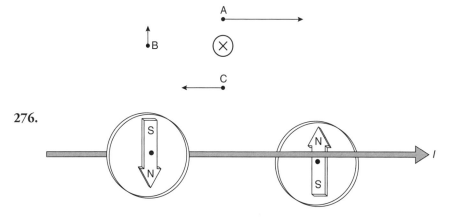

277. If you grasp a current-carrying wire with your right hand and the thumb pointing in the direction of the current, your fingers will curl around the wire in the direction of the magnetic field vectors.

278. Magnetic field is inversely proportional to the perpendicular distance from the wire.

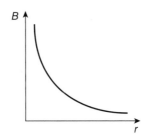

279. There are actually several different ways to use this right-hand rule. Here is the one that most of the current textbooks seem to be using. Hold your hand flat with your thumb at a right angle to your fingers.

- Your fingers should point in the direction of the magnetic field. There are usually lots of magnetic field lines, and you have lots of fingers.
- Your thumb will point in the direction of the current in the wire.
- Perpendicular to the palm of your hand is the direction of the force on the current-carrying wire. A nice way to think of this is that the direction of the force on the wire is the same direction in which you would push on something with the palm of your hand.

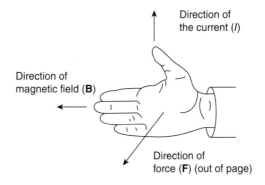

280. **(A)**

(B)

(C)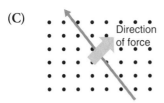

281. **(A)** The bottom wire produces a magnetic field out of the page, in the region of the top wire, which causes a force directed downward toward the lower wire.

(B) According to Newton's third law, the force is equal in strength but opposite in direction.

(C) The magnetic field is zero. The bottom wire produces an out-of-the-page magnetic field between the wires, while the top wire produces an into-the-page magnetic field. At the midpoint between the wires, the fields are equal in strength but opposite in direction. They cancel out.

282. **(A)** The bottom wire produces a magnetic field into the page, in the region of the top wire, which causes a force directed upward toward the top of the page and away from the lower wire.

(B) According to Newton's third law, the force is equal in strength but opposite in direction.

(C) The magnetic field from the top wire at the midpoint is into the page:
$$B_{\text{top wire}} = \frac{\mu_0}{2\pi} \frac{I}{\frac{d}{2}} = \frac{\mu_0}{\pi} \frac{I}{d}.$$

The magnetic field from the bottom wire at the midpoint is also into the page: $B_{\text{bottom wire}} = \frac{\mu_0}{2\pi} \frac{2I}{\frac{d}{2}} = \frac{\mu_0}{\pi} \frac{2I}{d}.$

Because the fields are in the same direction, the net magnetic field is also into the page:
$$B_{\text{bottom wire}} + B_{\text{top wire}} = \frac{\mu_0}{\pi} \frac{3I}{d}.$$

(D) The magnetic force on the top current-carrying wire is due to the magnetic field of the lower wire:

$$F_M = I_{\text{top}} l B_{\text{bottom}} = I_{\text{top}} l \left(\frac{\mu_0}{2\pi} \frac{I_{\text{bottom}}}{d} \right) = Il \left(\frac{\mu_0}{2\pi} \frac{2I}{d} \right).$$ Therefore, the force per unit length of wire is $\frac{F_M}{l} = \left(\frac{\mu_0}{\pi} \frac{I^2}{d} \right).$

283. **(A)** Upward. See figure.

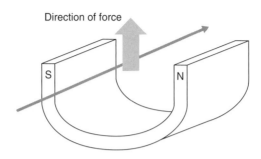

Direction of force

(B) The force will be doubled: $F_M = Il(\sin\theta)B$.

(C) The force is zero because the current in the bottom wire of the circuit and the magnetic field are parallel ($\sin\theta = 0$).

284. There are several different ways to perform this experiment. Here is one example:

(A) Equipment: A long thin strip of aluminum foil, batteries, switches, wires with alligator clips, and a magnet with north and south ends marked

(B)

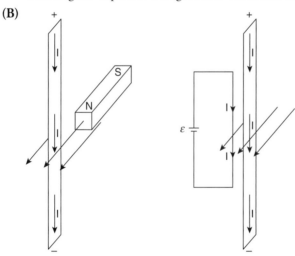

(C) Procedure:
1. Suspend the thin strip of aluminum foil vertically; connect the top end to the positive end of a battery and the bottom end to the negative end with a switch.
2. Place the north end of the magnet close to the strip so the magnetic field passes perpendicular to the strip, as shown in the figure.
3. Close the switch and observe the direction in which the aluminum strip is deflected. This indicates the direction of the force on the current in the strip. Be sure to open the switch to stop the current, as

you have produced a short circuit, which will produce a great deal of heat and drain the battery.
4. Now create another short circuit with a long wire, switch, and battery.
5. Suspend the wire parallel to the strip of foil.
6. Turn on both switches and observe the deflection of the foil. (Remember to open the switches to stop the current, as you have produced two short circuits, which will produce a great deal of heat and drain the battery.)
7. Using this method, you can investigate the effects of the magnetic field produced by a current-carrying wire by comparing it to the magnet's effect on the foil strip.

285. Gravitational fields point inward toward the mass, causing an attractive force inward on all other masses.

Electric fields radiate outward from positives and inward toward negatives. Positive charges experience a force in the direction of the field, while negative charges experience forces opposite to the field.

Magnetic fields are produced by moving charges according to the right-hand rule. Only moving charges experience magnetic forces. Moving positive charges experience a force perpendicular to the magnetic field, consistent with the right-hand rule. Moving negative charges receive a force in the opposite direction to what is predicted by the right-hand rule.

286. There are actually several different ways to use the right-hand rule. Here is the one that most of the current textbooks seem to be using. Hold your hand flat with your thumb at a right angle to your fingers.

- Your fingers should point in the direction of the magnetic field. There are usually lots of magnetic field lines, and you have lots of fingers.
- Your thumb will point in the direction of the velocity of the charged particle.
- Perpendicular to the palm of your hand is the direction of the force on a positive moving particle. (Negative moving particles will receive a force in the opposite direction.) A nice way to think of this is that the direction of the force on the particle is the same direction in which you would push on something with the palm of your hand.

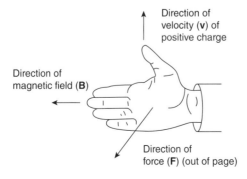

287. The directions of the forces are shown in the figure.

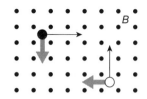

(A) $F_M = qv(\sin\theta)B = evB = 3.2 \times 10^{-13}$ N.

(B) The force is the same magnitude: 3.2×10^{-13} N.

288. (A) Out of the page.

(B) The force is zero because the velocity and field are parallel: $F_M = qv(\sin\theta)B = 0$.

(C) The force is out of the page. Since part of the velocity is perpendicular and part is parallel to the field, the electron will travel in a helical path circling along the magnetic field lines toward the bottom of the page.

289. Note that the neutron travels along a straight path. All positive particles curve toward the top of the page. The proton, electron, and positron all have the same charge, velocity, and magnetic force acting on them. Therefore, the lightest particles will curve with a tighter radius. The positron and electron have the same radius of curvature. The alpha particle has twice the charge of the proton, which means it will receive twice the force. However, it is also four times the mass of the proton. Therefore, it will have a greater momentum per magnetic force and will curve less.

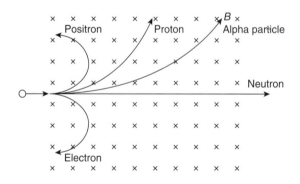

290. **(A)** The fields are in opposite directions at point P. Remember to convert your distance to meters.

$$B_{1A} = \frac{\mu_0}{2\pi} \frac{I}{r} = 5 \times 10^{-6} \ T \text{ into the page.}$$

$$B_{2A} = \frac{\mu_0}{2\pi} \frac{I}{r} = 8 \times 10^{-6} \ T \text{ out of the page.}$$

$$B_{net} = 3 \times 10^{-6} \ T \text{ out of the page.}$$

(B) The force is zero because the proton is not moving!

(C) $F_M = qv(\sin\theta)B = evB = 1.7 \times 10^{-18}$ N toward the top of the page.

291. **(A)** The magnetic field is downward at the origin because the magnetic field from the 2A wire dominates. By the right-hand rule, the force on the proton will be into the page.

(B) The fields are in opposite directions to the right of 8 cm and will cancel out where the fields have the same magnitude. This is satisfied at 12 cm:

$$B = \frac{\mu_0}{2\pi} \frac{I}{r} = \frac{\mu_0}{2\pi} \frac{2I}{2r}.$$

292. The bottom right and top left wires will produce magnetic fields directed upward to the right. The other two wires produce magnetic fields directed downward and to the right. The net sum of all the four individual magnetic fields is horizontally to the right.

293. **(A)** The magnetic field is to the left. By the right-hand rule, the force is out of the page.

(B) Into the page, by Newton's third law.

(C) The force is proportional to the velocity; therefore, the force is doubled. The direction is the same.

294. **(A)** Remember that the magnetic force is the centripetal force:

$$F_M = qv(\sin\theta)B = ma_C = m\frac{v^2}{r}$$

$$r = \frac{mv}{qB}$$

(B) Rearranging our equation, we get $mv = rqB$.

(C) Rearranging our equation, we get $\frac{q}{m} = \frac{v}{rB}$.

(D) Remember that the velocity of an object in circular motion is the circumference divided by the time period:

$$v = \frac{2\pi r}{T}$$

$$T = \frac{2\pi r}{v} = \frac{2\pi}{v}\left(\frac{mv}{qB}\right) = \frac{2\pi m}{qB}$$

Note that the time period is not dependent on the velocity of the charged particle.

Questions 295–301

295. The field should be uniform and constant, filling all the space between the plates. Due to the polarity of the battery, the top plate is positive. Therefore, the electric field is directed downward.

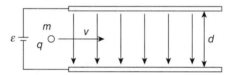

296. $E = \dfrac{\Delta V}{\Delta r} = \dfrac{\varepsilon}{d}$

297. The particle is positive and receives a constant electric force downward. This force will accelerate downward toward the bottom plate in a parabolic path.

298. We need a magnetic force upward. Using the right-hand rule, the magnetic field must point into the page.

299. The electric and magnetic forces must cancel out: $F_E = F_M$

$$Eq = \frac{\varepsilon q}{d} = qvB$$

$$B = \frac{\varepsilon}{dv}$$

300. The electric force remains the same, but the upward magnetic force increases because it is proportional to the velocity. Therefore, the particle will curve upward toward the top plate. $Eq < q(2v)B$.

301. Because the charge cancels out of the equation, the negative particle will travel straight through without curving: $Eq = qvB$.

302. **(A)** The magnetic force on the negative charge is to the left. We need an electric field to the left.

(B) The electric force on the electron is to the left. We need a magnetic field out of the page.

(C) The magnetic force is out of the page. We need an electric field into the page.

Questions 303–305

303. Use conservation of energy: $\Delta U = Q\Delta V = \Delta K = \dfrac{mv_2^2}{2} - \dfrac{mv_1^2}{2}$

$$\dfrac{mv_2^2}{2} = Q\Delta V + \dfrac{mv_1^2}{2}$$

$$v_2 = \sqrt{\dfrac{2Q\Delta V + mv_1^2}{m}}$$

304. Remember that the electric and magnetic forces must cancel out. $F_E = F_M$

$$Eq = \dfrac{\varepsilon q}{d} = qv_2 B$$

$$v_2 = \dfrac{\varepsilon}{dB}$$

305. Remember that the magnetic force is the centripetal force.

$$F_M = qv(\sin\theta)B = Qv_2 B = ma_C = m\dfrac{v_2^2}{r} = m\dfrac{v_2^2}{\left(\dfrac{D}{2}\right)}$$

$$D = \dfrac{2mv_2}{QB}$$

306. Diamagnetic materials have no unpaired electrons in their electron shells. These materials produce a weak, repulsive magnetic field when placed in an external magnetic field. When the external magnetic field is removed, the internal repulsive field disappears. Carbon, copper, and gold are diamagnetic.

Paramagnetic materials have at least one unpaired electron in their electron shells. These materials produce a weak, attractive magnetic field when placed in an external magnetic field. When the external magnetic field is removed, the internal attractive magnetic field disappears. Sodium, oxygen, and aluminum are paramagnetic.

Ferromagnetic materials have multiple unpaired electrons in their electron shells that easily produce a strong net magnetic field in the material. They have naturally forming magnetic domains inside the material. These domains can be aligned to produce a permanent magnet. Iron, cobalt, and nickel are ferromagnetic.

307. Only ferromagnetic materials have magnetic domains. Magnetic domains are areas where groups of atomic magnetic fields are aligned in the same direction.

308. When exposed to an external magnetic field, ferromagnetic domains align with the external field, producing a very strong magnetic field in the same direction as the external field. When the external magnetic field is removed, the domains can remain aligned, forming a permanent magnet.

309. A diamagnetic fluid will be weakly repelled by a strong magnet. A paramagnetic material will be weakly attracted to a strong magnet.

310. Magnetic flux occurs when magnetic field lines pass through an area.

311. $\Phi_B = B(\cos\theta)A$. To change the magnetic flux, all we need to do is change one of the variables in the equation. We can increase or decrease the magnetic field strength, change the angle between the area and the magnetic field, or change the area that is exposed to the magnetic field.

312. (A) There are many answers to this one. All we need to do is change one of the variables in the magnetic flux equation. Here are two examples: move the loop to the left, out of the magnetic field, or rotate the loop about a diameter.

(B) There are many answers. Here are two: move the magnet away from the loop, or drop the magnet through the loop.

(C) Here are two examples: rotate the magnet about its long axis as shown in the figure, or move both the magnet and the loop in the same direction at the same time, keeping the same distance between them, so there is no relative motion between them.

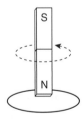

313. (A) $\Phi_B = B(\cos\theta)A$. Change the area by collapsing the loop of wire so it has a smaller cross-sectional area. Change the angle between the field and the loop by rotating the loop about a diameter. Change the field strength.

(B) The loop could be rotated clockwise or counterclockwise; alternatively, just move the loop to the right in the field without exiting the field.

(C) By Lenz's law, the induced magnetic field inside the loop will oppose the change in magnetic flux. Since the field into the page is getting stronger, the current will be counterclockwise to produce an opposing field out of the page.

314. **(A)** There is a change in magnetic flux only at locations 2 and 4.

(B) The flux is increasing into the page at location 2. Therefore, the current is counterclockwise to produce an out-of-the-page field to oppose the increase. At location 4, the flux is decreasing. Therefore, the current is clockwise to produce an into-the-page field to oppose the decrease in the flux.

315. **(A)** $\varepsilon = Blv = Byv$

(B) $I = \dfrac{\Delta V}{R} = \dfrac{Byv}{R}$

(C) The magnetic flux is increasing out of the page. Therefore, the current will be clockwise to produce a magnetic field into the page to oppose the increase in flux. Thus, the current will be upward through the resistor.

316. A microphone has a flexible membrane with a coil of wire attached that vibrates when sound waves strike the membrane. The coil sits near a magnet. The vibration causes changes in the magnetic flux that produce oscillating currents in the coil, which are then transmitted to a speaker to produce sound.

An electric generator consists of loops of wire attached to a rotating shaft. Turning the shaft swings the coils of wire past magnets. This produces a change in flux by changing the angle between the magnetic field vectors and the coils. The current induced in the wires is then transmitted from the generator to power electrical devices.

AP-Style Multiple-Choice Questions

317. **(C)** The original velocity is toward the top of the page. The force on the proton is out of the page. Therefore, by the right-hand rule, the magnetic field is directed to the left.

318. **(C)** The original velocity of the electron is toward the bottom of the page. The force on the electron is to the right. Therefore, by the right-hand rule, the magnetic field is out of the page. Remember that the electron is negative and receives a force opposite to what positive receives!

Questions 319 and 320

319. **(B)** $F_C = F_M = qvB = ma_C = m\dfrac{v^2}{r}$

$$r = \dfrac{mv}{qB} = 5 \times 10^{-3} \text{ m}$$

312 > Answers

320. **(D)** $r = \dfrac{mv}{qB}$. The mass of the electron is smaller; therefore, the radius is smaller as well. The charge is opposite, so the electric force is in the opposite direction.

321. **(A and D)** Be careful! Make sure you are paying attention to which of these is a magnetic field and which is an electric field. Electric forces are along the axis of the field. E-Fields push positive charges in the direction of the field. Negative charges are pushed in the opposite direction of the E-Field. Magnetic forces abide by the right-hand rule. Only moving charges experience forces from B-Fields.

322. **(B)** Newton's third law.

323. **(D)** It is not possible to separate a north pole from a south pole. All magnets are dipoles. When you break a magnet in half, you get two weaker magnets. If they stayed the same magnitude as the original, we would be violating conservation of energy.

324. **(C)** The dipoles in ferromagnetic materials align with the external magnetic field and amplify it. The dipoles in a diamagnetic material do not align with the external magnetic field.

325. **(B)** Use the right-hand rule for magnetic fields around current-carrying wires.

326. **(C)** By the right-hand rule, the magnetic force on the proton is to the left. Originally, the electric and magnetic forces were equal. Since the velocity has increased, the magnetic force is now larger than the electric force that is to the right.

Questions 327 and 328

327. **(A)** The magnetic field, due to the current in the wire near the electron, is into the page. By the right-hand rule, the negative electron will receive a force to the left due to the current. By Newton's third law, the force on the wire will be equal and opposite to the right.

328. **(A and C)** $F_M = qvB = (qv)\dfrac{\mu_0}{2\pi}\dfrac{I}{r}$

329. **(D)** The diaphragm vibrates back and forth along the axis of the magnet, changing the magnetic field strength through the coil area.

AP-Style Free-Response Questions

330. **(a)** The capacitor is connected to the battery so the left plate is at a higher potential than the right plate. This will accelerate positive ions to the right and hinder electrons from crossing between the plates.

(b) i.
$$K = \Delta U$$
$$\frac{1}{2} m_N v^2 = \Delta V q = \varepsilon e$$
$$v = \sqrt{\frac{2\varepsilon e}{m_N}}$$

ii. This must be a circular path, as shown in the figure.

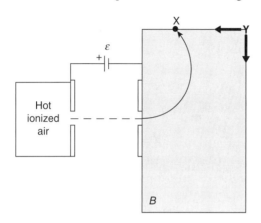

iii.
$$\sum F = ma = m_N a_C = m_N \frac{v^2}{r}$$
$$F_M = qvB = evB = m_N \frac{v^2}{r}$$
$$r = \frac{m_N}{eB} v = \frac{m_N}{eB}\sqrt{\frac{2\varepsilon e}{m_N}} = \frac{1}{B}\sqrt{\frac{2\varepsilon m_N}{e}}$$

(c) The radius goes up by a factor of $\sqrt{2}$. A reasonable answer range is shown in the figure above.

(d) The velocity of the doubly ionized nitrogen will be increased by a factor of $\sqrt{2}$ passing through the capacitor. Inside the magnetic field, the radius of motion is decreased by a factor of $\sqrt{2}$.

331. (a) The force of gravity of the electron is much smaller than the electric and magnetic forces. This is due to the mass of an electron being many times smaller than its charge and the universal gravitational constant being many times smaller than Coulomb's law and magnetic constants.

(b) i. The electric force and magnetic force cancel out. In the free body diagram, the forces must be equal, and the magnetic force must be upward, as shown in the figure.

$$F_E = F_M$$

ii.
$$F_E = F_M$$
$$Eq = \frac{\Delta V}{\Delta r} q = \frac{\Delta V}{d} q = qvB$$
$$\Delta V = vBd = 6{,}000 \text{ V}$$

iii. The bottom plate. The magnetic force on the electron is upward. Therefore, the electric force must be downward to cancel out the magnetic force. This means the electric field is directed upward, and the lower plate must be a higher potential.

iv.
$$U_C = \frac{1}{2}CV^2 = \frac{1}{2}\left(\frac{\varepsilon_0 A}{d}\right)V^2 = 8.0 \times 10^{-7} \text{ J}$$

(c) i. The electric force cancels out the magnetic force, and the charge drops out of the equations.
$$F_E = F_M$$
$$Eq = qvB$$
$$E = vB$$

ii. The magnetic force will be greater than the electric force. The path of the positron should be curving downward, as shown in the figure.

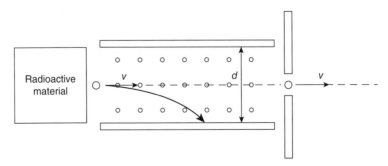

(d) i. The moving charge is experiencing a magnetic force from the current-carrying wire:

$$F_M = qvB_{\text{wire}} = qv\left(\frac{\mu_0}{2\pi}\frac{I}{r}\right)$$

The potential difference to resistance ratio is simply the current in the wire:

$$I = \frac{\varepsilon}{R} = \frac{F_M 2\pi r}{qv\mu_0} = 0.41 \text{ A}$$

ii. The current is to the clockwise around the circuit. The magnetic field from the top wire is out of the page in the vicinity of the electron. The magnetic force vector on the electron is directed upward toward the top of the page. The force vector arrow should be drawn upward.

Chapter 6: Geometric and Physical Optics

Skill-Building Questions

332. A wave is a disturbance in a medium that transports energy through that medium.

333. Differences: Longitudinal waves vibrate parallel to the direction of wave propagation, while transverse waves vibrate perpendicular to the direction of wave propagation. Transverse waves can be polarized, but longitudinal waves cannot. All electromagnetic (EM) waves are transverse.

Similarities: Both waves transport energy and have a frequency, time period, and amplitude. Both exhibit interference, diffraction, refraction, reflection, and Doppler effect.

334. Here are several different representations of longitudinal waves:

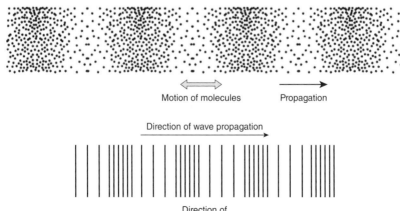

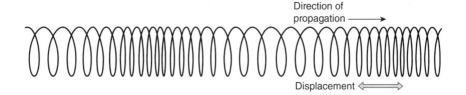

Here is a representation of a transverse wave and an EM wave.

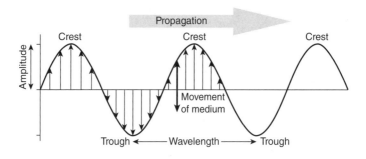

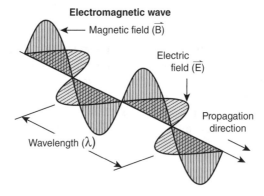

335. Differences: The biggest difference is that mechanical waves require a physical medium to transport through. Electromagnetic waves do not require a physical medium as they create their own electric/magnetic field and are said to be self-propagating. Thus, EM waves travel through vacuums, but mechanical waves do not. Mechanical waves can vibrate as longitudinal waves or transverse waves, depending on how they are created and the physical medium they pass through. All EM waves are transverse, with both a transverse electric field wave and a transverse magnetic field wave perpendicular to each other. Electromagnetic waves all travel at the speed of light c in a vacuum and "slow down" when moving through mediums. Mechanical waves travel much slower, at speeds determined by the medium, generally traveling fastest in solids and slowest in gases.

Similarities: Both exhibit the classic wave properties of interference, diffraction, refraction, reflection, and Doppler effect.

336. Note the wave equation is shown as a cos function, but it can also be a sin function.

Equation 1: E = electric field strength as a function of time t
A = amplitude of the wave measured from the undisturbed medium position, which is usually the midpoint between the crest and the trough
f = frequency of the oscillating disturbance
T = time period of the oscillating disturbance
t = time

Equation 2: B = magnetic field strength as a function of location
A = amplitude of the wave measured from the undisturbed medium position, which is usually the midpoint between the crest and the trough
x = location
λ = wavelength

337. (A) $E = A\sin\left(\dfrac{2\pi t}{T}\right) = 0.3\sin\left(\dfrac{2\pi t}{2.4}\right) = 0.3\sin(2.6t)$

(B) $E = A\cos\left(\dfrac{2\pi x}{\lambda}\right) = 60\cos\left(\dfrac{2\pi x}{1.5}\right) = 60\cos(4.2x)$

(C) $B = A\cos\left(\dfrac{2\pi x}{\lambda}\right) = 0.60\cos\left(\dfrac{2\pi x}{2}\right) = 0.60\cos(\pi x)$

338. (A) $B = 4.0\cos\left(\dfrac{\pi x}{2}\right)$. This is a cosine wave with an amplitude of 4.0 T and wavelength of 4 m.

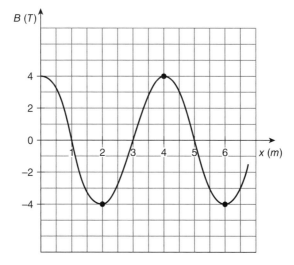

(B) $E = 2\sin(1.05t)$. This is a sine wave with an amplitude of 2 N/C and wavelength of 6 s.

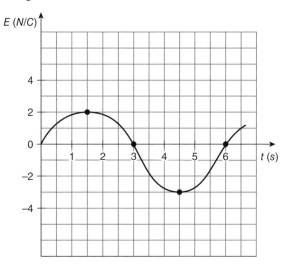

339. When waves occupy the same space, the net amplitude is the sum of the amplitudes of the separate waves.

340. Constructive interference: When two waves overlap in phase (crest to crest), the net amplitude increases.

Destructive interference: When two waves overlap out of phase (crest to trough), the net amplitude decreases.

341. Here are six of the interference patterns seen as the two waves overlap and then pass by each other.

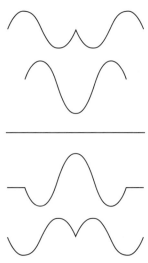

342. The Doppler effect occurs when the source of the wave and the observer of the wave have a relative velocity between them. When the two are moving apart, the EM wave will have a longer wavelength and lower frequency (red shift). When the two are moving toward each other, the EM wave will have a shorter wavelength and higher frequency (blue shift). Examples of technology that use the Doppler effect are Doppler weather radar and police radar guns. Astronomers use red and blue shift to measure how fast stars are moving away from or toward the earth.

343. Electromagnetic waves are a vibration in electric and magnetic fields. The changing electric field produces the magnetic field, and the changing magnetic field produces the electric field. Thus, an EM wave can propagate itself through a vacuum.

344. Electromagnetic waves can be polarized, and only transverse waves can be polarized.

345. The rope has a vibration axis. When the rope axis is aligned with the axis of the picket fence, the wave passes through unchanged. However, when the axes are perpendicular, the wave cannot pass through the fence.

346. (A) Nonpolarized light is vibrating along all the axes equally. Thinking of this in terms of x-y components, this means that 50 percent is vibrating along the x-axis, and 50 percent is vibrating along the y-axis. Therefore, a vertical polarizing filter will only let through the 50 percent of the light vibrating along the y-axis; it will block the light vibrating along the x-axis.

(B) The first filter blocks horizontal axis light. The second lets through horizontal axis light, but there is none left. Therefore, no light passes through the second filter.

(C) The first filter lets vertical axis light through. The second filter lets angled axis light through. The vertically polarized light has a component of light along the angled axis. Therefore, a component of the vertical light will pass through the second filter.

347. Radio, microwave, infrared, visible, ultraviolet, X-ray, gamma ray

348. Similarities: Both have the same method of propagation. Both are EM waves that travel at the speed of light.

Differences: The only difference is the wavelength/frequency and energy of the wave. Radio waves have longer wavelengths, lower frequencies, and less energy than X-rays.

349. All EM waves travel at $c = 3.0 \times 10^8$ m/s in a vacuum.

350. 400 nm (violet light) to 700 nm (red light)

351. Using the equation, $v = f\lambda$, we get a frequency range of approximately 4.3×10^{14} Hz (red light) to 7.5×10^{14} Hz (violet light).

352. Wave fronts show where the wave crests and troughs are located. Rays indicate in which direction the wave is traveling. Rays are always perpendicular to wave fronts.

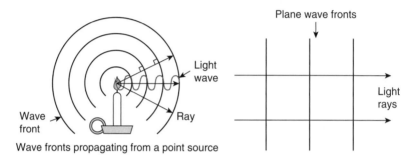

353. Huygens' principle states that each point on a wave front is the source of a wavelet that moves outward from the original wave front. The superposition of all the individual wavelets creates the new wave front.

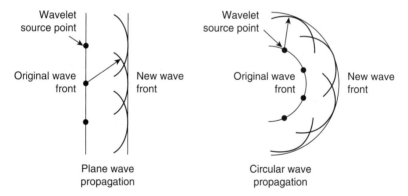

354. **(A)** The portion of the wave that hits the boundary on the right side will be reflected or absorbed. The left half of the wave will pass by the boundary. The wavelets that are produced at the edge of the boundary will cause the wave to curve around the edge of the boundary.

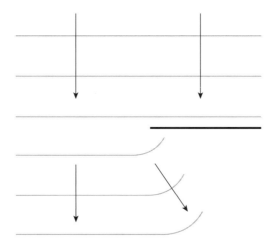

(B) Sound waves have a long wavelength, and wavelets near the boundary will cause a pronounced diffraction effect. As the wavelength gets smaller than the boundary itself, as in visible light waves, the diffraction effect is less pronounced because the wavelets near the boundary are small.

(C) The point source model shows that when the opening is comparable in size to the wavelength, there will be a pronounced diffraction (bending of the wave front) around the corners of the opening.

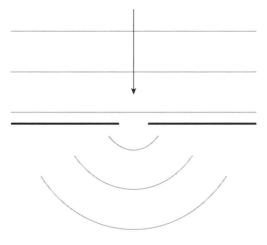

(D) When the opening is much larger than the wavelength, the point source model shows that only the edges of the wave show any bending or diffraction. The wavelets in the center continue the propagation of the plane wave.

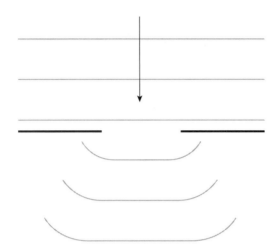

(E) The point source model shows that the two slits will produce two separate curved wave fronts passing through the openings. These two new wave fronts will interfere with each other, creating a pattern of constructive and destructive interference.

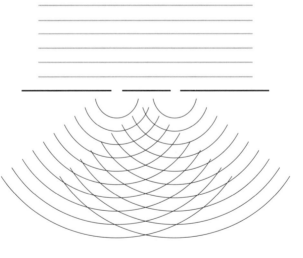

(F) The speed of light in the new medium is slower. Drawing the Huygens' wavelets, we can see that all parts of the wave will slow down at the same time. This means the wave front will not change direction but will simply travel straight through the glass with a shorter wavelength.

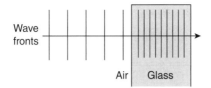

(G) Entering the new medium, the wave slows down. Drawing the wavelets in the new medium with a shorter wavelength, we see that the wave front changes direction, and the wave ray bends toward the normal line to the surface.

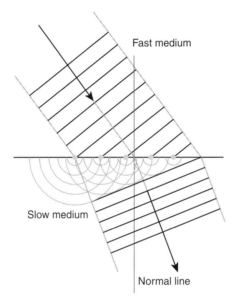

(H) Drawing wavelets for the wave that reflects off the surface, we see that the wave maintains the same speed and wavelength. However, the reflected wave front turns and travels off in a new direction. The angle of the incoming wave front measured to the normal is equal to the angle of the outbound reflected wave front measured to the normal line.

355. The light and dark pattern of single-slit diffraction is described by this equation: $d\sin\theta = m\lambda$, where d is the slit or opening width, λ is the wavelength of the light, θ is the angle from the slit to the minima (dark spot) on the screen, and m is the minima number. Note that $m = 1$ is the first minima to either side of the central maxima (bright spot), and $m = 2$ would be the second minima. Solving the equation for our situation, we get $\sin\theta = \dfrac{m\lambda}{W}$. Therefore, as λ goes up or W goes down, θ increases.

(A) As X increases, there will be no change to the pattern because nothing in the equation changes.

(B) As Y increases, nothing in the equation changes. However, since the screen is farther away from the barrier, there is a greater distance for the pattern to spread out before it hits the screen. Therefore, the pattern will become wider or more spread out, as shown in the figure.

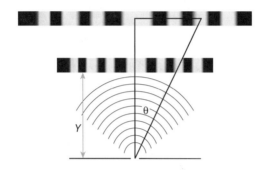

(C) When W increases, θ decreases. Therefore, the pattern becomes narrower and more tightly packed.

(D) Red light has a longer wavelength, which will increase θ, making the pattern spread out.

(E) Violet light has a shorter wavelength, which will decrease θ, making the pattern narrower.

356. This is a situation of single-slit diffraction: $d\sin\theta = m\lambda$. The deflection angle to the first-order minima is determined by the ratio of wavelength over opening width: $\sin\theta \propto \dfrac{\lambda}{d}$. The width of a door is roughly 1 m, which is similar in size to the wavelengths of sound. Therefore, the ratio $\dfrac{\lambda}{d} \approx 1$ and $\theta = \sin^{-1}\left(\dfrac{\lambda}{d}\right) \approx 90°$. This means the sound will diffract all the way around the corner and fill the entire room beyond. Light has a very small wavelength (400–700 nm). Therefore, the ratio $\dfrac{\lambda}{d} \approx 0$ and $\theta = \sin^{-1}\left(\dfrac{\lambda}{d}\right) \approx 0°$. This means the light will diffract very little as it passes through the doorway because the doorway is so much larger than the wavelength of light. Light actually does diffract through the door—just not very much.

357. The light and dark pattern of double-slit diffraction is described by this equation: $d\sin\theta = m\lambda$, where d is the distance between the two slits; λ is the wavelength of the light; θ is the angle from the slit to the maxima (bright spot) on the screen;

and m is the maxima number. Note that $m = 0$ is the central maxima right down the center along the line of symmetry, and $m = 1$ would be the first maxima to either side of the central maxima. Solving the equation for our situation, we get $\sin\theta = \dfrac{m\lambda}{Z}$. Therefore, as λ goes up or Z goes down, θ increases.

(A) As W decreases, there will be no change to the pattern because nothing in the equation changes.

(B) As X decreases, there will be no change to the pattern because nothing in the equation changes.

(C) As Y decreases, nothing in the equation changes. However, since the screen is closer to the barrier, there is less distance for the pattern to spread out before it hits the screen. Therefore, the pattern will become closer together and more tightly packed.

(D) As Z decreases, θ increases. Therefore, the pattern becomes wider and more spread out.

(E) Red light has a longer wavelength, which will increase θ, making the pattern spread out.

(F) Violet light has a shorter wavelength, which will decrease θ, making the pattern narrower.

(G) The maximas will be in the same locations but will be thinner, more point-like, and not so spread out.

358. A bright spot will occur whenever there is constructive interference. This happens when the waves from the two openings reach the screen in phase (crest-crest or trough-trough). This will always occur in the center along the line of symmetry, because both waves travel the same distance to reach that location. This also occurs when the path length difference (ΔL) to a point on the screen is a multiple of the wavelength: $\Delta L = m\lambda$. Therefore, there will be multiple locations of constructive interference corresponding to path length differences of $\Delta L = 1\lambda, 2\lambda, 3\lambda$, and so on.

Dark spots will occur where there is destructive interference. This happens when the waves from the two openings reach the screen out of phase (crest-trough). This occurs when the path length difference (ΔL) to a point on the screen is a half-multiple of the wavelength: $\Delta L = m\lambda$. Therefore, there will be multiple locations of destructive interference corresponding to path length differences of $\Delta L = \dfrac{1}{2}\lambda, \dfrac{3}{2}\lambda, \dfrac{5}{2}\lambda$, and so on.

359. Similarities: Both patterns are described by the equation $d\sin\theta = m\lambda$, where d is the distance between the two adjacent slits; λ is the wavelength of the light; θ is the angle from the slit to the maxima (bright spot) on the screen; and m is the maxima number. Therefore, the pattern spacing is the same.

Differences: Double-slit maxima are spread out, while multislit diffraction grating maxima are very tight and point-like, as shown in the figure.

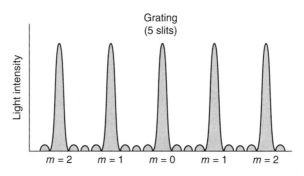

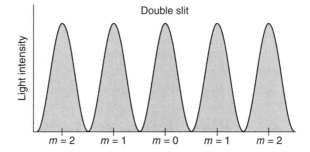

360. When light strikes a thin film, some light reflects off the top of the film. The rest travels through the film, and some of this reflects off the bottom of the film and back up through the film again. This transmitted light has traveled a distance equal to twice the thickness of the film. To produce constructive interference, the thickness will need to be ½λ so the path length difference is a whole multiple of the wavelength. This will cause the two waves to align in phase, creating constructive interference. The index of refraction of the film will change the wavelength of the light in the thin film. Thus, the thickness of the thin film depends on the wavelength of the light in the film itself, not in the air.

When light strikes a more optically dense medium, the reflected wave will be inverted. Inverted reflections are a phase change equivalent to the wave jumping ½λ. If the wave is inverted at both reflections, the effect cancels out. However, if only one inversion occurs, the thickness of the thin film will need to be only ¼λ thick because the interfering waves are already ½λ apart.

361. The law of reflection: The incoming angle = the reflected angle measured from the normal to the surface.

Answers ‹ 327

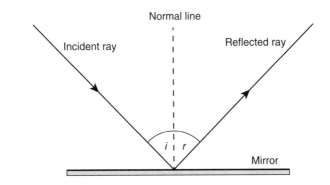

362.

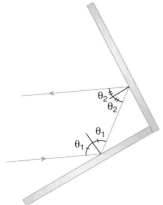

363.

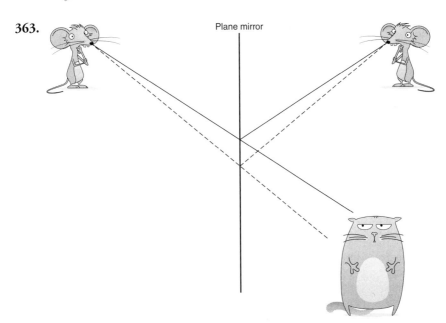

364. When light encounters a boundary, the light can be reflected, refracted (transmitted), or absorbed. The more similar the substances, the more the light will be refracted through the new medium. The less similar the substances, the more the light will reflect off the surface. The total amount of light energy that is reflected, refracted, and absorbed will equal the original energy of the light striking the surface.

365. The velocity of the light and wavelength will both decrease according to the equations: $n = \dfrac{c}{v} = \dfrac{\lambda_{vacuum}}{\lambda_{medium}}$. When light travels into a new medium only the frequency will remain the same.

366. $v_{vacuum} \gtrsim v_{air} > v_{water} > v_{glass}$

367. (A) $v = \dfrac{c}{n} = \dfrac{3 \times 10^8 \text{ m/s}}{1.33} = 2.26 \times 10^8$ m/s

(B) $\lambda_{water} = \dfrac{\lambda_{vacuum}}{n} = \dfrac{580 \text{ nm}}{1.33} = 436$ nm

(C) The frequency in water is the same as in air: $f = \dfrac{c}{\lambda} = \dfrac{3 \times 10^8 \text{ m/s}}{580 \text{ nm}} = 5.17 \times 10^{14}$ Hz

(D)

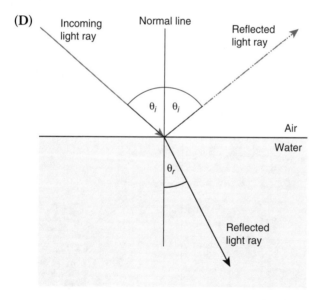

Questions 368–371

368.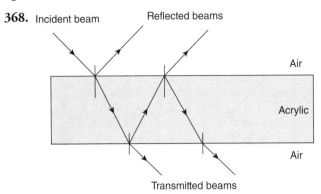

369. The dots on the left are brighter both above and below the acrylic block. Conservation of energy tells us that less and less of the light is making it through to the right dots.

370. Ray #1 must be striking the acrylic/air interface at an angle of incidence larger than the critical angle. This produces total internal reflection along the parallel top and bottom surfaces, as shown by the dashed line in the figure.

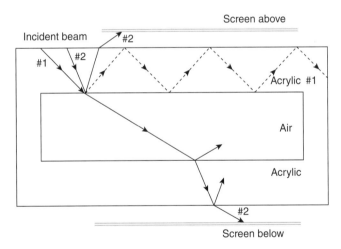

371. Ray #2 must be striking the acrylic/air interface at an angle less than the critical angle. There is a partial reflection and refraction at each surface. This will produce a dot on each screen, as shown by a solid line in the figure from question 370.

372. **(A)**

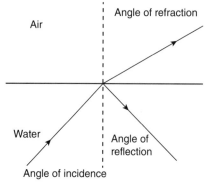

(B) $\theta_{reflection} = \theta_{incidence} = 45°$
$n_a \sin\theta_a = n_w \sin\theta_w$
$\sin\theta_a = 1.33\sin(45°)$
$\theta_a = 70°$

(C) $n_a \sin(90°) = n_w \sin\theta_{critical}$
$\sin(90°) = 1.33\sin\theta_{critical}$
$\theta_{critical} = 49°$

Therefore, an angle equal to 49° or larger will not enter the air.

(D) No! Traveling from air into water, the light ray will slow down and turn toward the normal. There is no critical angle. Therefore, some of the light will always enter the water from the air. Even when the angle of incidence from the air is 90 degrees, some of the light will enter the water at an angle of 49 degrees.

$$n_a \sin\theta_a = n_w \sin\theta_w$$
$$\sin(90°) = 1.33\sin\theta_w$$
$$\sin\theta_w = 0.75$$
$$\theta_w = 49°$$

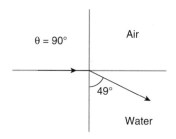

373. **(A)** Equipment: protractor, pencil, ruler, sheet of white paper

(B)

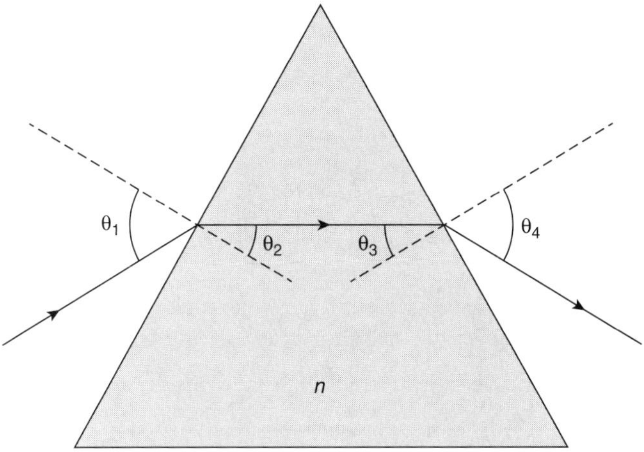

[Diagram of a triangular prism with index n. A ray enters the left face at angle θ_1 from the normal, refracts to angle θ_2 inside, travels through the prism, strikes the right face at angle θ_3 from the normal inside, and exits at angle θ_4.]

(C) Procedure:
1. Place the prism in the center of the sheet of paper.
2. Trace the outside of the prism with the pencil.
3. Direct the laser beam so it passes through the prism.
4. Trace the ray's path with the ruler before it enters the prism and after it exits, being sure to mark the entrance and exit locations.
5. Remove the prism. Mark the normal line at the incoming and outgoing surfaces. Draw the light ray's path through the prism.
6. Using the protractor, measure the angles of incidence and refraction for each incoming and outgoing ray.
7. Use Snell's law to calculate the index of refraction of the glass prism.
8. To reduce error, repeat for a wide variety of angles, and average the result.

374. **(A)** Use Snell's law and the first set of data:

$$n_a \sin\theta_a = n_s \sin\theta_s$$

$$\sin(10°) = n_s \sin(6°)$$

$$n_s = 1.66$$

Repeating this for the rest of the data in the table, we get an average of 1.76.

(B) We can see in the figure that the trend line clearly shows the data are not straight and cannot be a direct relationship.

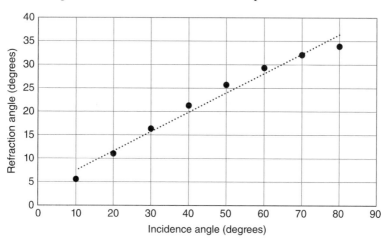

(C) Plotting the sine of the refracted angle as a function of the sine of the incident angle produces a straight line. This makes sense:

$$n_s \sin\theta_s = n_a \sin\theta_a = \sin\theta_a$$
$$n_s \sin\theta_s = \sin\theta_a$$
$$\sin\theta_s = \frac{1}{n_s}\sin\theta_a$$
$$y = mx + b$$

Comparing Snell's law with the equation of a line, we can determine what the slope represents. The slope of the line will equal the reciprocal of the index of refraction of the sapphire. The best fit line gives a slope of 0.565. Thus, $n_s = 1.77$.

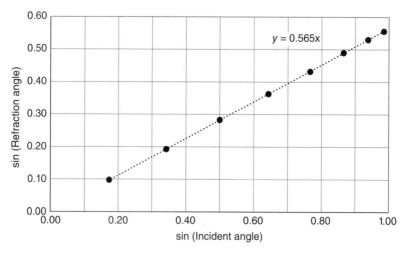

375. **(A)** The incident ray partially reflects and refracts at the surface. This allows for multiple exit points from the prism.

(B) Note that all angles with the normal are smaller inside the prism than the rays in the air. This is because the light is traveling more slowly in the prism than in the air.

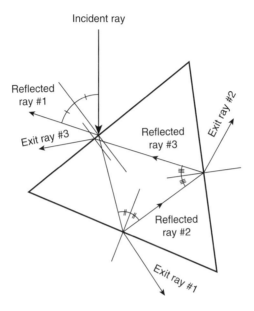

(C) Note that all the reflected angles in the preceding figure obey the law of reflection. The incoming and reflected rays are the same. This is shown with marks like the ones used in geometry class.

376. **(A)** Using Snell's law:

$$n_a \sin\theta_a = n_p \sin\alpha$$
$$\sin(30°) = 1.5\sin\alpha$$
$$\alpha = 19°$$

(B) Using some geometry, we can deduce that the angle beta is 41 degrees. (Use the idea that the internal angles inside any triangle add up to 180 degrees.) Using Snell's law, we see that the light ray does indeed exit the prism as shown in the figure:

$$n_p \sin\beta = n_a \sin\gamma$$
$$1.5\sin(41°) = \sin\gamma$$
$$\gamma = 80°$$

377. **(A)** Remember that at the critical angle, the refracted angle is 90°. Calculating the minimum index of refraction so light does not escape at a critical angle of 45 degrees is shown by the following equation:

$$n_a \sin(90°) = n_{glass} \sin\theta_{critical}$$
$$\sin(90°) = n_{glass} \sin(45°)$$
$$n_{glass} = 1.41$$

(B) In the left figure, the light is attempting to exit out of the glass into air but strikes the bottom of the prism beyond the critical angle. When the prism is lowered into water, the difference in the index of refraction between the glass and water is less than between the glass and air. Therefore, the light does not speed up as much going into the water as it did going into air. This reduces the bending of the light due to refraction. The light is no longer hitting the bottom of the prism beyond the critical angle. Therefore, the light is able to escape out the bottom of the glass into the water.

378.

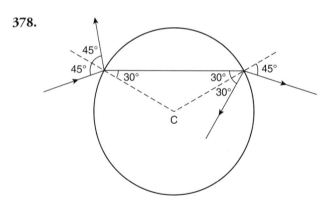

379. Fiber optics consist of a thin transparent tube. Light shines in one end. Because the fiber is so thin, the angle of incidence inside the tube is always beyond the critical angle, and the light cannot escape until it reaches the other end of the fiber.

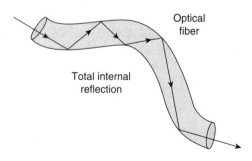

380. Both images can be seen. Real images are formed by light converging to a point and, thus, can be projected on a screen. Virtual images are formed by diverging light rays that appear to come from a specific point but have not. Virtual images cannot be projected on a screen.

381. The focal length of a spherical mirror is equal to half the radius of the mirror.

382. The focus, like the name implies, is the location where parallel light rays striking the mirror converge for concave mirrors and convex lenses. Parallel rays appear to diverge from the focus for convex mirrors and concave lenses.

383. Lenses A, C, and F are converging/convex lenses because they are thicker in the middle than at the edges. Lenses B, D, and E are diverging/concave lenses because they are thinner in the middle than at the edges. Mirror G is converging/concave. Mirror H is diverging/convex.

384. For converging/convex lenses:
1. A ray, parallel to the principal axis, refracts through the lens and passes through the focus on the opposite side of the lens.
2. A ray passing through the front focus and is refracted parallel to the principal axis on the other side of the lens.
3. A ray passes straight through the center of the lens, and its direction does not change.

For diverging/concave lenses:

1. A ray, parallel to the principal axis, refracts away from the principal axis as if it has come from the near-side focus.
2. A ray directed toward the far-side focus is refracted parallel to the principal axis.
3. A ray passes straight through the center of the lens, and its direction does not change.

336 > Answers

Questions 385–389

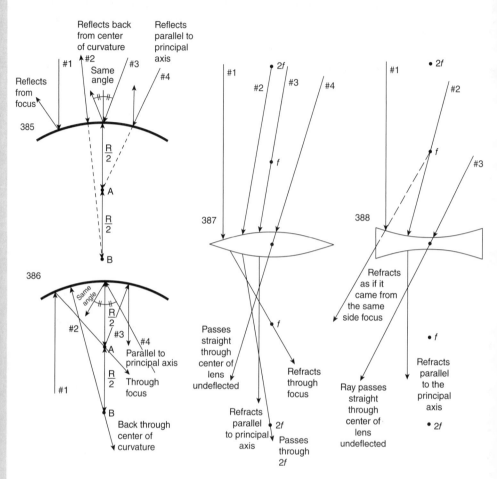

385. See figure.

386. See figure.

387. See figure.

388. See figure.

389. The reflection from the mirrors will remain exactly the same, as the law of reflection does not depend on the medium. The rays traveling through the lenses will refract less because there will be less of a velocity change when the light travels from water to lens.

390. The mirror's focal length will not change as it depends only on the curvature of the mirror. However, lens optics depend on both the shape of the lens and refraction. The refraction of the lens will be less for oil/lens than for air/lens because the light will not change speed as much. Less refraction means the light rays will bend less going through the oil/lens, and the focal length will be greater than for the air/lens configuration.

391. (A) Note that the rays are closer to the normal line inside the lenses because the speed of light is slower inside the lens than in air.

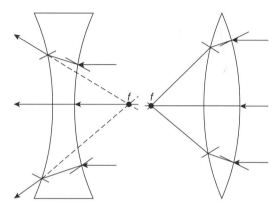

(B) Increasing the thickness of the lens will increase the angle of incidence at which the incoming ray strikes the lens. This increases the refraction at the surface and the deflection of the inbound rays but decreases the focal length of the lens.

392. (A-C)

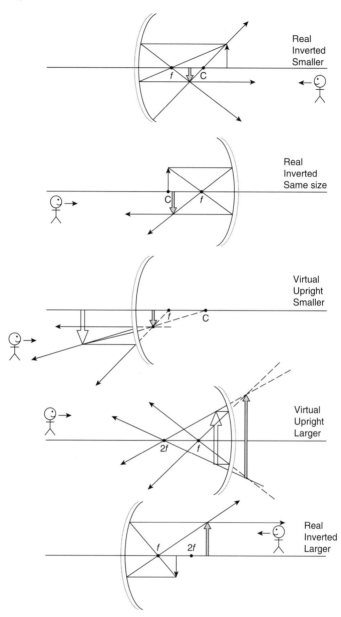

(D) Light rays converge to form real images. Light rays diverge from the location of a virtual image as if they came from that spot, even though they did not actually pass through that location.

393. (A-C) See figure.

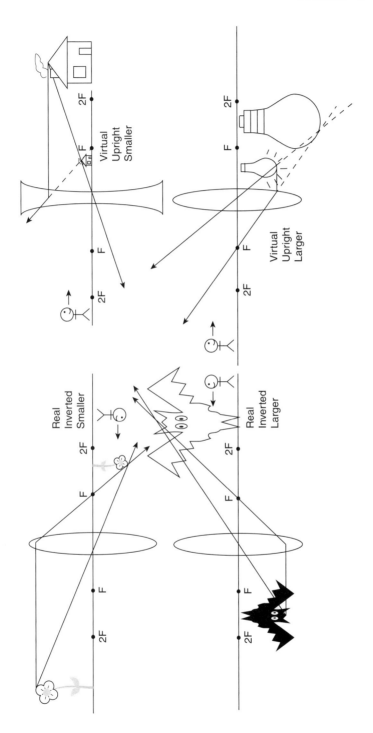

(D) Light rays converge to form real images. Light rays diverge from the location of a virtual image as if they came from that spot, even though they did not actually pass through that location.

(E) Since half of the light rays will still be passing through the lens unimpeded, the image will still form in the same spot with the same properties as before. However, the image will not be as bright because half of the rays from the flower are now blocked.

394.

Type of optical device	Object location	Focal length (+ or −)	Image location s_i (+ or −)	Real or virtual	Inverted or upright	Image magnification
Diverging lens	Between F and 2F	−	− Same side of lens	Virtual	Upright	Smaller
Convex lens	Beyond 2F	+	+ Between F and 2F	Real	Inverted	Smaller
Converging lens	Between F and 2F	+	+ Beyond 2F	Real	Inverted	Larger
Convex lens	Between F and the lens	+	− Same side of lens	Virtual	Upright	Larger
Converging mirror	Beyond 2F	+	+ Between F and 2F	Real	Inverted	Smaller
Concave mirror	Between F and 2F	+	+ Beyond 2F	Real	Inverted	Larger
Converging mirror	Between F and the mirror	+	− Behind the mirror	Virtual	Upright	Larger
Convex mirror	Between F and the mirror	−	− Behind the mirror	Virtual	Upright	Smaller

395. The light rays will converge to the focus. Therefore, the focal length is 20 cm. The radius will be twice the focal length, or 40 cm.

396. Since the image is inverted, the lens must be a converging lens with a positive focal length:

$$\frac{1}{f} = \frac{1}{s_i} + \frac{1}{s_o}$$

$$\frac{1}{10 \text{ cm}} = \frac{1}{30 \text{ cm}} + \frac{1}{s_o}$$

$$s_o = 15 \text{ cm}$$

$$M = \frac{-s_i}{s_o} = \frac{-30 \text{ cm}}{15 \text{ cm}} = -2$$

The image is twice as large as the object and inverted. The object is 15 cm from the lens.

397. Since the image is upright and larger, the mirror must be concave, with a positive focal length creating a virtual image. The image is upright, so the magnification must be positive.

$$M = +2.5 = \frac{-s_i}{s_o} = \frac{-s_i}{5 \text{ cm}}$$

$$s_i = -12.5 \text{ cm}$$

This makes sense because virtual images have a negative image distance.

$$\frac{1}{f} = \frac{1}{s_i} + \frac{1}{s_o}$$

$$\frac{1}{f} = \frac{1}{-12.5 \text{ cm}} + \frac{1}{5 \text{ cm}}$$

$$f = 8.3 \text{ cm}$$

The focal length is positive just as we expect for a concave mirror. The radius of curvature will be twice the focal length, 16.6 cm.

398. We need to be careful here! Diverging optical devices have a negative focal length, and we expect the image to have a negative image distance because it will be virtual. Since we have a diverging optical device, the magnification should be positive (upright) and less than one.

$$\frac{1}{f} = \frac{1}{s_i} + \frac{1}{s_o}$$

$$\frac{1}{-40 \text{ cm}} = \frac{1}{s_i} + \frac{1}{20 \text{ cm}}$$

$$s_i = -13.3 \text{ cm}$$

$$M = \frac{-s_i}{s_o} = \frac{-(-13.3 \text{ cm})}{20 \text{ cm}} = +0.67$$

399. Eyes produce a real image on the retina, so the focal length should be positive.

$$\frac{1}{f} = \frac{1}{s_i} + \frac{1}{s_o}$$

$$\frac{1}{f} = \frac{1}{1.7 \text{ cm}} + \frac{1}{30 \text{ cm}}$$

$$f = 1.6 \text{ cm}$$

400. Diverging optical devices have a negative focal length. The focal length of a mirror is half the radius of curvature; therefore, the focal length is −25 cm. Diverging optical devices only produce virtual images with negative image distances, so the distance to the image is −15 cm.

$$\frac{1}{f} = \frac{1}{s_i} + \frac{1}{s_o}$$

$$\frac{1}{-25 \text{ cm}} = \frac{1}{-15 \text{ cm}} + \frac{1}{s_o}$$

$$s_o = 37.5 \text{ cm}$$

$$M = \frac{-s_i}{s_o} = \frac{-(-15 \text{ cm})}{37.5 \text{ cm}} = 0.4$$

The magnification tells us that the image is upright and smaller than the object. This is what we would expect to get from a diverging optical device.

401. There are several ways to perform this lab. Here is one method.
  (A) Equipment: mirror, meter stick, candle, screen
  (B)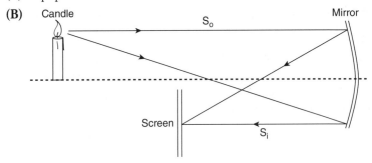
  (C) Procedure:
    1. Place the mirror, candle, and screen on a table, as shown in the figure.
    2. Move the screen until a crisp image forms.
    3. Measure the object distance from the mirror to the candle and the image distance from the mirror to the screen.
    4. Repeat this process for several data points.
    5. Graphing $\frac{1}{s_o}$ on the x-axis and $\frac{1}{s_i}$ on the y-axis, we should get a graph with a slope of −1 and an intercept that equals $\frac{1}{f}$. (You can also use the mirror equation to calculate the focal length for each set of data and average, but the AP test usually asks you to graph a straight line graph to find what you are looking for.)
  (D) No! A convex mirror will not produce a real image on the screen.

402. There are several ways to perform this lab. Here is one method.
- **(A)** Equipment: lens, meter stick, candle, screen
- **(B)**

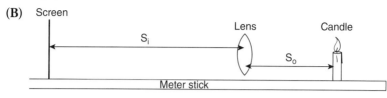

- **(C)** Procedure:
    1. Place the lens, candle, and screen on a table, as shown in the figure.
    2. Move the screen until a crisp image forms.
    3. Measure the object distance from the lens to the candle and the image distance from the lens to the screen.
    4. Repeat this process for several data points.
    5. Graphing $\frac{1}{s_o}$ on the x-axis and $\frac{1}{s_i}$ on the y-axis, we should get a graph with a slope of −1 and an intercept that equals $\frac{1}{f}$.

    (Note: Another easy way to find the focal length is to take the lens outside on a sunny day. Produce a point of light with the rays from the sun. The distance from the lens to the point of light will be the focal length.)
- **(D)** No! A concave lens will not produce a real image on the screen.

Questions 403 and 404

403. Graphing $\frac{1}{s_o}$ on the x-axis and $\frac{1}{s_i}$ on the y-axis, we should get a graph with a straight line with a slope of −1.

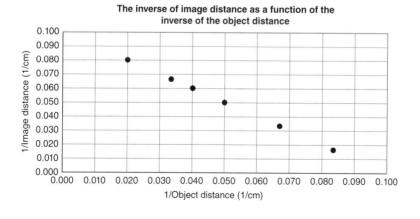

404. Rearranging the lens equation and comparing it to the equation of a line, we can see that the intercept of the line equals $\frac{1}{f}$.

$$y = mx + b$$
$$\frac{1}{s_i} = (-1)\frac{1}{s_o} + \frac{1}{f}$$

The intercept equals 0.1 (1/cm). Therefore, the focal length is 10 cm.

405. **(A)** The crystal ball is behaving like a convex/converging lens because it is inverting the image.

(B)

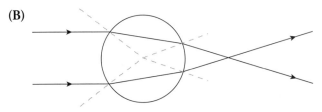

406. **(A)** The image formed on the retina is real. Therefore, the lens must be convex/converging.

(B) A concave/diverging lens is needed to spread the rays outward so that the eye can focus them on the retina.

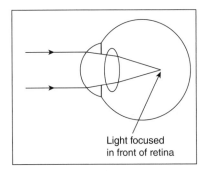 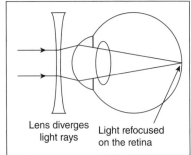

(C) A converging/convex lens is needed to converge the light rays before they enter the eye. The lens in the eye converges the light the rest of the way to focus the image on the retina.

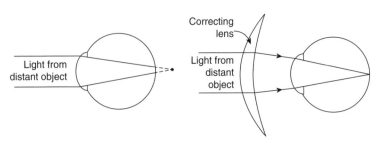

AP-Style Multiple-Choice Questions

407. (C) The law of reflection is not influenced by the water. Snell's law of refraction depends on the indices of refraction of the two materials. The speed of light changes less going from water to lens then going from air to lens. This means there will be less refraction in water, making the focal length larger.

408. (B) Electromagnetic waves are transverse, with both the E- and B-fields oscillating perpendicular to the direction of motion and each other.

409. (A and B) $\sin\theta = \dfrac{m\lambda}{d}$. Increasing the wavelength of the laser (λ) and/or decreasing the hair width (d) will both increase the angle of the pattern.

410. (D) The point source model shows us that the larger the wavelength, the greater the bending of the wave around the corner. We can also see this in the equation for diffraction: $\sin\theta = \dfrac{m\lambda}{d}$. As the wavelength gets smaller, the diffraction bending angle also gets smaller. Visible light has a very small wavelength and only bends a tiny amount around corners.

411. (C) The amplitude is about 3 mV/m, and the time period is 0.5 ns. Using the wave equation, we get

$$E = A\cos\left(\dfrac{2\pi t}{T}\right) = 3\cos\left(\dfrac{2\pi t}{0.5}\right) = 3\cos(4\pi t) = 3\cos(12.6t)$$

412. (A) The image is upright and smaller. This means the image is virtual, the image distance is negative, and the mirror must be diverging/convex.

$$M = \dfrac{1}{2} = \dfrac{-s_i}{s_o} = \dfrac{-s_i}{12\text{ cm}}$$

$$s_i = -6\text{ cm}$$

$$\dfrac{1}{f} = \dfrac{1}{s_i} + \dfrac{1}{s_o}$$

$$\dfrac{1}{f} = \dfrac{1}{-6\text{ cm}} + \dfrac{1}{12\text{ cm}}$$

$$f = -12\text{ cm}$$

413. (A) The frequency remains the same. The wavelength of light is directly proportional to the speed of light in the substance. Light travels faster in water than through glass; therefore, $\lambda_w > \lambda_g$.

414. (D) The lens equation can be rearranged to produce a straight line:

$$\frac{1}{s_i} = (-1)\frac{1}{s_o} + \frac{1}{f}$$

$$y = mx + b$$

Thus, if we plot $\frac{1}{s_o}$ on the *x*-axis and $\frac{1}{s_i}$ on the *y*-axis, we should get a graph with a slope of -1 and an intercept of $\frac{1}{f}$. The image distance is $x_2 - x_1$, and the object distance is $x_0 - x_1$.

415. (B) When the angle of the prism decreases, the right side of the prism becomes more vertical, and the angle of incidence with the normal becomes smaller. This creates less refraction and the distance (x) increases. Consider the extreme case when θ becomes zero. Then the angle of incidence is zero, and there is no refraction at all. The beam will pass straight through the "prism" because it has become flat like a window, and the distance (x) becomes infinite.

416. (B and D) Drawing the normal lines will make the paths more evident. The light ray should bend toward the normal when entering glass and bend away from the normal when entering air. Note that answer choice D shows light entering/exiting the prism along the normal and with total internal reflection inside the prism.

417. (A) *Be careful of units!* Convert the object distance from 30 cm to 300 mm.

$$\frac{1}{f} = \frac{1}{s_i} + \frac{1}{s_o}$$

$$\frac{1}{16 \text{ mm}} = \frac{1}{s_i} + \frac{1}{300 \text{ mm}}$$

$$s_i = 16.9 \text{ mm}$$

This means the image is 3.1 mm in front of the retina, which is at a distance of 20 mm. We need a diverging lens to move the image back to the retina. Diverging lenses are concave.

AP-Style Free-Response Questions

418. (a) First, we need to find the object and image distance from the data in the table. Knowing that the lens is located at 50 cm and using the first set of data, we get the following:

$$s_o = 50 \text{ cm} - 0 \text{ cm} = 50 \text{ cm}$$

$$s_i = 71 \text{ cm} - 50 \text{ cm} = 21 \text{ cm}$$

The focal length can be calculated using the lens equation:

$$\frac{1}{f} = \frac{1}{s_i} + \frac{1}{s_o}$$

$$\frac{1}{f} = \frac{1}{50 \text{ cm}} + \frac{1}{21 \text{ cm}} = \frac{0.068}{\text{cm}}$$

$$f = 15 \text{ cm}$$

(We can confirm this data point by calculating the focal point with the other data points in the table. The last two columns will be used for part b.)

Candle location (cm)	Screen location (cm)	s_o (1/cm)	s_i (1/cm)	$1/s_o$ (1/cm)	$1/s_i$ (1/cm)
0	71	50	21	0.020	0.048
10	74	40	24	0.025	0.042
20	80	30	30	0.033	0.033
25	88	25	38	0.040	0.026
29	100	21	50	0.048	0.020

(b) The lens equation can be rearranged to produce a straight line:

$$\frac{1}{s_i} = (-1)\frac{1}{s_o} + \frac{1}{f}$$

$$y = mx + b$$

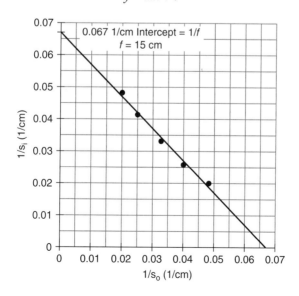

Thus, if we plot $\frac{1}{s_o}$ on the x-axis and $\frac{1}{s_i}$ on the y-axis, we should get a graph with a slope of −1 and an intercept of $\frac{1}{f}$. From our graph, the intercept is 0.067 1/cm, which gives us a focal length of 15 cm (the same as part a).

(c) i. An upright image will be virtual. The object will need to be between the lens and the focal point. Sketches will vary, and image locations will depend on where the object is placed between the focal point and the lens.

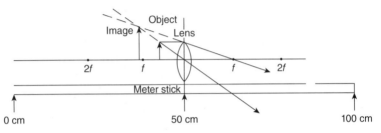

ii. Do not agree! Virtual images can be seen but cannot be projected on a screen. A simple demonstration would be to produce a real image and project it on a screen. Then produce a virtual image and show that the image cannot be made to show up on the screen.

419. (a) The left figure rays should all be parallel and pointing downward, with the angle of refraction larger than angle of incidence. The right figure should also show the waves traveling in a more downward direction. All wave fronts should be parallel, and the wavelength between the wave fronts should be larger in medium #2.

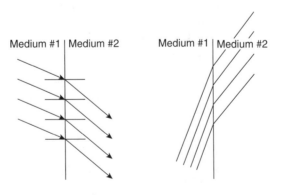

(b) i. The sketched waves should bend around the boundary and overlap in circular paths that maintain the same wavelength.

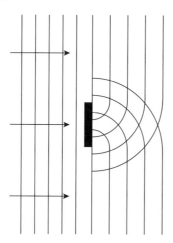

ii. The point source model says that every point on the wave is the source of a new wavelet that propagates outward. For the plane waves approaching the barrier, the sum of all the wavelets produces another plane wave in front of the last one. The barrier blocks the center wavelets. The plane waves on either side of the boundary continue forward, but the wavelets on the end of the blocked wave produce curved waves that propagate inward to fill the central area beyond the boundary. These curved waves will overlap and form constructive interference, where crests meet crests and troughs meet troughs. They will create destructive interference where crests meet troughs.

(c) i. There are several ways to draw this. The interference pattern shown here is marked to show the constructive interference points. An amplitude function could also be used to represent the interference pattern. The key points of the drawing are that it is symmetrical along the central axis and that there is a central maximum. We do not have enough information to calculate the exact locations of the constructive and destructive interference. What is important is that they alternate in an evenly spaced pattern and are marked with C and D as seen in the figure.

350 > Answers

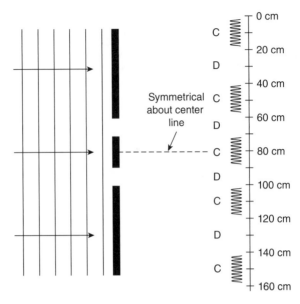

ii. The pattern will get wider and spread out from the central maximum. As the distance between the wave fronts is increased, the wavelength gets larger. This means the pattern will spread out as the angle θ gets larger: $d\sin\theta = m\lambda$.

420. **(a)** i. The ray must turn away from the normal to the right.

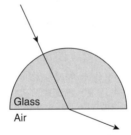

ii. The speed of light in air is approximately 3×10^8 m/s. In glass, it travels more slowly. When the ray exits the glass and enters air, the light speeds up. This causes the wavelength to get larger, and the wave turns away from the normal line.

(b) Writing Snell's law for this situation and lining it up with the equation for a line, we see that if we plot $\sin\theta_a$ on the y-axis and $\sin\theta_g$ on the x-axis, we should get a straight line with a slope of n_g and an intercept of zero:

$$n_g \sin\theta_g = n_a \sin\theta_a = \sin\theta_a$$
$$\sin\theta_a = n_g \sin\theta_g$$
$$y = mx + b$$

(c) Equipment: protractor, pencil, ruler, sheet of white paper
Procedure:
1. Place the semicircular prism in the center of the sheet of paper.
2. Trace the outside of the prism with the pencil. Mark a spot in the center of the flat surface of the prism where the laser beam is to exit each time.
3. Direct the laser beam perpendicular through the curved surface of the prism toward the spot in the center of the flat surface. Use the ruler to verify that the beam goes straight through the curved surface of the prism without bending.
4. Trace the ray's path with the ruler before it enters the prism's curved surface and after it exits the prism at the mark on the flat side.
5. Repeat for a wide variety of angles, and number each incoming and outgoing ray so you know which one is paired up with which when you measure the angles later.
6. Remove the prism. Mark the normal line at the flat surface. Extend the light rays to the normal line/flat surface intersection.
7. Measure the angles of incidence and refraction for each incoming and outgoing ray.
8. Build a table of data and plot $\sin\theta_a$ on the y-axis and $\sin\theta_g$ on the x-axis. The slope will be the index of refraction of the glass.

(d) i. Traveling along path A, the light reflects and refracts at the top surface. Some light escapes the top. The rest reflects toward the left surface, where some of the light refracts and escapes. Traveling along path B, the light hits the top surface at an angle larger than the critical angle, and none can escape the top surface. All the light reflects toward the left surface, where some refracts and escapes.

ii. and iii.

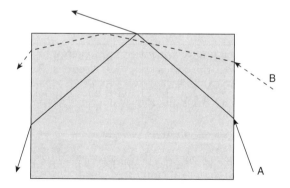

iv. Path B will produce the brightest beam, exiting on the left side. None of the path B beam exited the top of the prism. By conservation of energy, there will be more light left to exit the prism on the left side.

Chapter 7: Quantum, Atomic, and Nuclear Physics

Skill-Building Questions

421. **(A)** The speed of light. Relativity says that no matter what, everyone measures the same speed of light.

    (B) The speed of light: 3×10^8 m/s.

422. The laws of physics are the same in all inertial reference frames. Thus, the laws of physics are the same for everyone. The speed of light is always 3×10^8 m/s in all inertial reference frames. Thus, everyone measures the same speed of light, even if the source or the observer is moving.

423. We start to see the effects of time dilation and length contraction when moving about one tenth the speed of light and faster. Thus, we usually don't see the effects of relativity in our everyday lives.

424. A person moving away from the Earth in a spaceship with a speed of 0.5 c will not agree with a person on earth about the time it takes to get to the nearest star. Nor will they agree on the distance to the star. Events that happen on the spaceship seem to happen at different times, as measured by the person in the spaceship and the observer on Earth.

425. The photoelectric effect is the phenomenon where light shining on an object can cause electrons to be ejected from (knocked off) the object. Electrons are held by the atoms in the material (work function). The light must have an energy at least as large or higher than this work function for an electron to be ejected. This is the experiment that showed the particle properties of light, because light of too low a frequency cannot eject electrons no matter how bright the light is. Einstein coined the term *photon* to describe the new particle nature of light. Photons of light have energy that is dependent on their frequency: $E = hf$.

426. Ultraviolet photons have enough energy to dislodge electrons from a negatively charged object that has excess electrons. Positively charged objects lack electrons. The only way to discharge a positively charged object is to add electrons or remove protons. A UV photon cannot do either of these.

427. Figure A shows photons of light striking potassium, which has a work function of 2.29 eV. This means it takes 2.29 eV of energy to knock an electron off potassium. Photons of light with less than this value will not be able to eject an electron because photons interact with electrons like particles on a one-to-one basis. Note that 700 nm does not have enough energy to produce photoelectrons. The energy of photoelectrons is measured by forcing them to travel through a potential difference that tries to stop them. The stopping potential is increased

until none of the electrons can reach the other plate, and the photocurrent drops to zero, as shown in Figure B. The maximum energy of the ejected electrons is $K_{max} = U_{electric} = q\Delta V = e\Delta V = hf - \Phi_{work\ function}$.

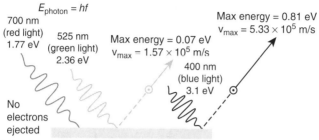

Fig. A

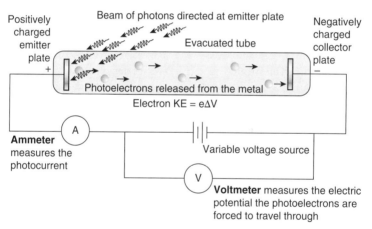

Fig. B

428. Photon energy: The energy of a single particle of light. Photon energy depends only on the frequency of the wave: $E = hf$.

Stopping potential: This is the electric potential used to stop the photoelectrons to determine their kinetic energy: $K_{max} = U_{electric} = q\Delta V = e\Delta V$.

Threshold/cutoff frequency: This is the frequency below which photons of light no longer have enough energy to produce photoelectrons (knock electrons off the metal): $hf_T = \Phi_{work\ function}$.

Work function: This is the minimum amount of energy needed to remove an electron from a material. Think of it as how much energy the atom expends to hold on to the outermost electron.

429. There are two phenomena in the photoelectric experiment that support the particle model of light. First, both dim light and very bright (intense) light of the same wavelength and frequency eject electrons with exactly the same maximum kinetic energy. This does not make sense according to the wave model of light, which says that brighter light has a larger amplitude and should eject electrons with more kinetic energy. The particle model of light solves this problem. Dim light is simply a stream of a few photons. Bright light is a stream of many photons. Dim light will eject fewer photoelectrons than bright light because it has fewer photons. However, the ejected electrons from both dim light and bright light have exactly the same kinetic energy. This is because all the photons have the same frequency and, therefore, the same energy.

Second, for each photosensitive material tested, the energy of the ejected photoelectrons was directly related to the frequency of the light striking the surface. Even stranger, below a certain threshold/cutoff frequency, no electrons will be ejected no matter how intense or bright the light is. According to the wave model, it should be possible to use intense but low frequency light to remove electrons from the material. The particle model solves this problem as well. Light consists of particles (photons) of light whose energy is directly proportional to the frequency of the wave: $E = hf$. Low frequency light cannot eject electrons because it simply does not have the energy to do so. The higher the frequency of light, the more energy the photoelectrons will have: $K_{max} = hf - \Phi_{work\ function}$.

430. The energy of all electromagnetic (EM) waves depends only on the frequency of the light: $E = hf$.

431.

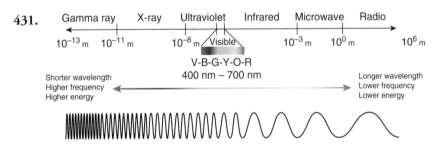

432. An electron volt is a unit of energy convenient when in the atomic scale: 1 eV = 1.6×10^{-19} C.

433. **(A)** At the threshold frequency, the kinetic energy of the photoelectrons is zero:

$$K_{max} = 0 = hf_T - \Phi$$

$f_{T\ sodium} = 5.8 \times 10^{14}$ Hz, $f_{T\ iron} = 1.1 \times 10^{15}$ Hz, $\Phi_{cesium} = 1.9$ eV.

(B)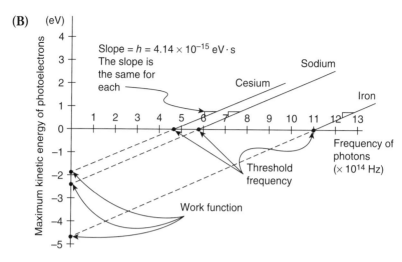

(C) $K_{max} = hf - \Phi = \dfrac{hc}{\lambda} - \Phi$

$K_{max} = \dfrac{1{,}240 \text{ eV} \cdot \text{nm}}{200 \text{ nm}} - 1.9 \text{ eV} = 4.3 \text{ eV}$

434. **(A)** Wave property: Only waves exhibit interference.

(B) Particle property: Light behaves like a particle by colliding with an electron, thus imparting momentum to the electron and knocking it from the atom.

(C) Wave property: Interference is a wave property.

(D) Particle property: Collisions and momentum are particle properties.

(E) Wave property: Diffraction, the bending of waves around boundaries, is a wave property.

(F) Wave property: The Doppler effect is a property of waves.

(G) Particle property: Solar cells, or photovoltaic cells, are a practical application of the photoelectric effect, which demonstrates the particle nature of light.

435. No! Electromagnetic waves/photons of light do not have a net charge and are immune to external electric and magnetic fields.

436. Photons are massless particles that have both energy and momentum.

437. Photon energy: $E = hf$. Photon momentum: $p = \dfrac{h}{\lambda}$.

Remember that for a wave, $v = f\lambda$, and the speed of a light wave is c.

The relationship between photon energy and momentum: $E = hf = \dfrac{hc}{\lambda} = pc$.

438. **(A)** The X-ray will lose energy and momentum. Conservation of momentum must be obeyed in the collision. Since the electron gained momentum in the collision, the photon must lose momentum. Also, the electron gains kinetic energy in the collision. Therefore, the photon must lose energy. Losing momentum and energy means the frequency of the X-ray has decreased: $E = hf = pc$.

(B) The electron gains both kinetic energy and momentum to the right from the collision with the X-ray.

439. **(A)** The X-ray will lose energy and momentum. Conservation of momentum must be obeyed in the collision. Since the electron gained momentum in the collision, the photon must lose momentum. Also, the electron gains kinetic energy in the collision. Therefore, the photon must lose energy. Losing momentum and energy means the frequency of the X-ray has decreased: $E = hf = pc$.

(B) Elastic! Electrons have no internal structure. Therefore, there are no dissipative internal forces to turn the kinetic energy into thermal energy.

(C) Energy: $(E_{photon})_{initial} = (E_{photon} + E_{electron})_{final}$

$$(hf)_{initial} = (hf + K_{electron})_{final}$$

$$(hf)_{initial} = \left(hf + \frac{1}{2}m_{electron}v^2\right)_{final}$$

Momentum in the x-direction:

$$(p_{photon})_{initial} = (p_{photon} + p_{electron})_{final}$$

$$\left(\frac{h}{\lambda}\right)_{initial} = \left(\frac{h}{\lambda}\cos\beta + m_{electron}v\cos\alpha\right)_{final}$$

Momentum in the y-direction:

$$0 = (p_{photon} + p_{electron})_{final}$$

$$0 = \left(-\frac{h}{\lambda}\sin\beta + m_{electron}v\sin\alpha\right)_{final}$$

440. The Rutherford model of the atom consisted of a nucleus in the center of the atom where all the positive charge resided. This area is approximately 10^{-14} m in diameter. The diameter of the atom as a whole was approximately 10^{-10} m in diameter. This means the nucleus was much smaller than the atom itself. The rest of this atomic space around the nucleus was where the tiny elections resided. At the time, nobody had any idea what the electrons were doing in this big open area of the atom.

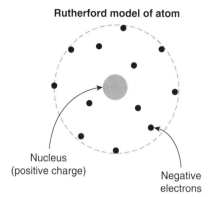

Rutherford model of atom

441. Accelerating charged particles radiate energy in the form of EM waves. Orbiting electrons would lose energy by radiation and crash into the nucleus, destroying the atom. So it cannot be correct.

442. Bohr "fixed" the planetary model of electron orbits by assuming stable orbital locations where the electrons would not radiate energy.

443. To move from one stable state to another required the emission or absorption of a photon of light. Moving up to a higher energy state required the absorption of a photon. Falling to a lower energy state required the emission of a photon. The photon energy always equals the difference in energy of the two stable energy states.

Questions 444–450

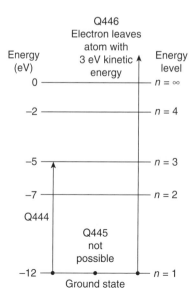

444. The 7-eV photon will be absorbed taking the electron from the ground state to the $n = 3$ energy level, the second excited state, as shown in the figure. The atom now has 7 eV more energy.

445. The 9 eV photon is not absorbed by an electron in the ground state because there is no stable energy stage at −3 eV. There is no energy change in the atom.

446. The 15-eV photon has enough energy to knock the electron free of the atom with 3 eV of kinetic energy. The atom is ionized.

447. Starting in the $n = 4$ energy level, the electron can follow any path that eventually takes the electron back to the ground state. All the possible transitions are shown in the figure. Each of these transitions will be accompanied by a photon emission, with energy equal to the difference in energies of the two states.

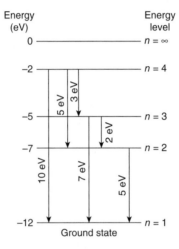

448. **(A)** $\Delta E = hf = \dfrac{hc}{\lambda}$. The frequency is directly related to the energy change in the transition. Therefore, the greater the ΔE, the greater the f of the emission. The frequency ranking from highest to lowest is B = D > A > C.

(B) The wavelength is inversely related to the energy change in the transition. Therefore, the smaller the ΔE, the greater the λ of the emission. The wavelength ranking is opposite that of frequency: C > A > B = D.

(C) Since photons are emitted from the atom during these transitions, the atom is losing energy, as shown in the figure.

449. **(A)** $\Delta E = hf = \dfrac{hc}{\lambda}$. The frequency is directly related to the energy change in the transition. Therefore, the greater the ΔE, the greater the f of the absorbed photon. The frequency ranking from highest to lowest is A > B = C.

(B) The wavelength is inversely related to the energy change in the transition. Therefore, the smaller the ΔE, the greater the λ of the absorbed photon. The wavelength ranking is B = C > A.

(C) Since photons are being absorbed during these transitions, the atom is gaining energy.

450. Note that the energy change is −3 eV. This means the atom is losing energy and a photon will be emitted by the atom. The absolute value of the ΔE is taken in the following equation so the answers come out positive.

$$\Delta E = E_{final} - E_{initial} = hf = \frac{hc}{\lambda}$$

$$\left| -5 \text{ eV} - (-2 \text{ eV}) \right| = (4.14 \times 10^{-15} \text{ eV} \cdot \text{s}) f = \frac{(1,240 \text{ eV} \cdot \text{nm})}{\lambda}$$

$$f = 7.26 \times 10^{14} \text{ Hz}, \lambda = 413 \text{ nm}$$

This is visible! It falls in the visible spectrum (400–700 nm) on the violet end of the spectrum.

451. (A)

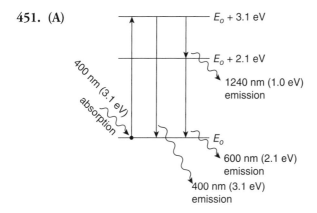

The problem does not tell us the starting position of the electron or its initial energy. So let's just label it at E_0. Using the equation $\Delta E = hf = \frac{hc}{\lambda}$, we can calculate the change in energy of the electrons in the atom associated with the photons:

$$\Delta E = \frac{(1,240 \text{ eV} \cdot \text{nm})}{\lambda}$$

The wavelengths of 400 nm and 600 nm give us energies of 3.1 eV and 2.1 eV, respectively.

(B) Yes! The figure shows that when the electron is in the $E_0 + 3.1$ eV energy level, it can fall to the $E_0 + 2.1$ eV energy level. This will produce a 1,240 nm/1.0 eV photon that is not in the visible spectrum, so it will not be seen.

Questions 452–455

452. The energy of the atom increases during the pumping phase as electrons are moved to higher energy states. The atom loses energy during the lasing phase as the stored energy of the electron in the E_2 energy level is released in the form of a photon.

453. Energy gain during the pumping phase is equal to the energy change during the two transitions:

$$E_{gain} = (E_3 - E_1) + (E_2 - E_3) = E_2 - E_1$$

454. The energy of the photon equals the energy difference of the two energy states. This will be a negative number because the atom is ejecting a photon, so it is shown as an absolute value:

$$E_{photon} = |E_1 - E_2|$$

455. $E_{photon\ \#1} = E_{photon\ \#2} + E_{photon\ \#3}$

456. Scientists have shown that massless EM waves have particle properties. Louis de Broglie suggested that particles with mass might also have wave properties. This turned out to be true and is one of the foundations of quantum mechanics. The wave nature of particles is named after de Broglie. The de Broglie wavelength of a particle with mass depends on the particle's momentum (mv): $\lambda = \dfrac{h}{p}$.

457. Both are inverse relationships, as seen in the equation for the de Broglie wavelength: $\lambda = \dfrac{h}{p} = \dfrac{h}{mv}$.

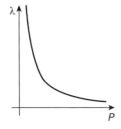
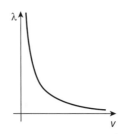

458. Shooting small particles, like electrons, through a double slit or single slit will produce an interference pattern, which is a wave property. Directing a beam of electrons at a crystal produces a diffraction pattern similar to that produced by X-rays. Electrons having wave properties explains why Bohr's stable electron energy states exist (see the answer to question 460). It also explains the electron energy shells that we all learned about in chemistry.

459. Electron beams can produce de Broglie wavelengths on the same order of magnitude and smaller than the atomic crystal spacing. A baseball will generally have a wavelength many magnitudes smaller than the atomic spacing in a crystal. Baseball wavelengths are much too small to produce an interference pattern. (This is the same reason light does not produce interference patterns when passing through two doorways, but sound waves do.) In addition, the baseball itself is larger than the atomic spacing of the crystal, so it is not physically possible for the baseball to interact with a single atom of the crystal at a time.

460. The electron only exists in "orbits" that are whole number multiples of the electron wavelength. These orbits are "stable" because the electron exhibits constructive interference in this location. At other locations, the electron wavelength is either too long or too short to produce constructive interference. The electron cannot exist in these locations because there is destructive interference, and the orbit is not stable.

461. **(A)** This is an interference pattern. Therefore, electrons are exhibiting wave properties (wave-particle duality).

(B) When the accelerating voltage increases, the electron velocity increases: $\Delta U = q\Delta V = \frac{1}{2}mv^2$. When electron velocity increases, the de Broglie wavelength of the electron decreases: $\lambda = \frac{h}{p} = \frac{h}{mv}$. When wavelength decreases, the pattern gets tighter, and the angle of the maxima decreases: $d\sin\theta = m\lambda$.

(C) The interference pattern will become more spread out as the distance d decreases: $\sin\theta = \frac{m\lambda}{d}$

462. A particle's wave function describes the wave nature of the particle. It shows us the probability of finding the particle at a particular location. The larger the amplitude of the wave function, positive or negative, the higher the likelihood of the particle being found at that location.

463. (A and B)

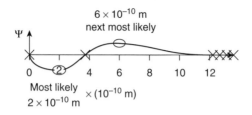

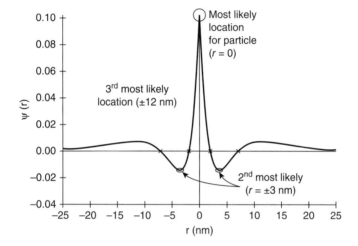

464. **(A)** This is a capacitor. The top plate has a higher electric potential. The electric field is directed straight down toward the bottom plate. The electric field is uniform in strength and direction between the plates except for at the edges of the capacitor.

(B) Ray C must not have a charge because it is not affected by the electric field. Ray B curves in the direction of the electric field. Therefore, it must be receiving a force from the field and must be positively charged. Ray A must be negative as it receives a force in the opposite direction of the electric field and curves upward.

(C) Ray C could be a neutron or a gamma ray because neither has an electric charge. Ray B could be a proton or an alpha particle. Ray A could be an electron. Ray A curves more tightly, which could be accounted for by an electron having less mass and therefore more acceleration than an alpha particle or a proton as in the case of ray B. However, we do not know the speed of the particles exiting the box. So it is hard to draw any conclusions by comparing the curve radius of rays A and B.

465.

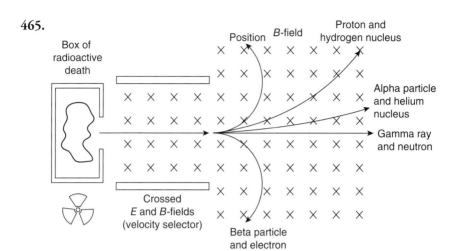

466. The atomic mass unit is a convenient unit to use when dealing with atomic-size particles. It is equal to 1.66×10^{-27} kg = 931 MeV/c^2. This is 1/12 of the mass of a carbon-12 nucleus.

467. Note that several versions of the particle symbol are shown in the table. Several particles in the table have different names but are otherwise identical.

Particle	Symbol	Mass (u)	Mass (kg)	Charge (e)	Charge (C)
Proton	p, $_1^1p$, $_1^1H$	1.0073 u	1.673×10^{-27}	$+e$	1.6×10^{-19}
Electron	β, e, $_{-1}^0\beta$, $_{-1}^0e$	0.0005 u	9.11×10^{-31}	$-e$	-1.6×10^{-19}
Neutron	n, $_0^1n$	1.0087 u	1.675×10^{-27}	0	0
Alpha particle	α, $_2^4\alpha$, $_2^4He$	4.0012 u	6.644×10^{-27}	$+2e$	3.2×10^{-19}
Beta particle	β, e, $_{-1}^0\beta$, $_{-1}^0e$	0.0005 u	9.11×10^{-31}	$-e$	-1.6×10^{-19}
Gamma ray	γ, $_0^0\gamma$	0	0	0	0
Positron	β^+, e^+, $_1^0\beta$, $_1^0e$	0.0005 u	9.11×10^{-31}	$+e$	1.6×10^{-19}
Hydrogen nucleus	p, $_1^1p$, $_1^1H$	1.0073 u	1.673×10^{-27}	$+e$	1.6×10^{-19}
Helium nucleus	α, $_2^4\alpha$, $_2^4He$	4.0012 u	6.644×10^{-27}	$+2e$	3.2×10^{-19}

468. Atoms consist of a nucleus composed of protons and neutrons. The nucleus is about 10^{-14} m in diameter. Electrons exist outside the nucleus in an electron cloud that is about 10^{-10} m in diameter.

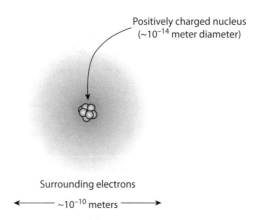

469. The electrons are held in the atom by the electrostatic attraction between the positive protons in the nucleus and the negative electrons.

470. The protons are repelled by the electric force. Gravity pulls the nucleons together but is too weak to hold the nucleus together. The nuclear strong force is stronger than electrostatic repulsion of the protons. The strong force attracts all the nucleons together, making the nucleus stable.

471. Isotopes are nuclei with the same number of protons (same element) but a different number of neutrons.

472. Not all atoms are stable. The nuclear strong force is a short-range force that holds adjacent nucleons together. As the nucleus grows, electrostatic repulsion of the protons can cause decay, as in uranium. Most isotopes are unstable. There are 15 known isotopes of carbon. Only C-12 and C-13 are stable. C-14 has a half-life of 5,730 years. None of the isotopes of uranium are stable.

473. Isotopes have the same number of protons and electrons. Since the electrons are the same, all isotopes have the same chemical properties and chemical behavior. Isotopes do not have the same nuclear properties because some are stable and others are not.

474. The mass of a nucleus is always less than the sum total of the individual masses of the protons and neutrons that comprise it. The difference in mass between a nucleus and its nucleon parts is the mass defect. Some of the nucleon mass is converted into energy that binds the nucleus together: $E = mc^2$. To break a nucleus apart, energy equal to the mass defect must be added to the nucleus to reproduce the missing mass. The mass defect and binding energy is an "energy" explanation for nuclear stability. The nuclear strong force is a "forces" explanation for why the nucleus stays together.

475. Similarities: Both nuclear binding energy and ionization energy refer to the input energy required to separate the particles from a nucleus or an electron from an atom. To separate a nucleon from a nucleus requires energy input to recreate the missing mass of the mass defect. To separate an electron from an atom, energy input is needed to overcome the attraction of the nucleus and the electron.

Differences: The nuclear binding energy is much larger than the ionization energy.

476. $\Delta m = (2m_{\text{neutron}} + 2m_{\text{proton}}) - m_{\text{He}}$

477. Find the total mass of the 26 protons and 30 neutrons that make up this nucleus:

$$m_{\text{nucleons}} = 26m_{\text{proton}} + 30m_{\text{neutron}}$$

$$m_{\text{nucleons}} = 26(1.673 \times 10^{-27} \text{ kg}) + 30(1.675 \times 10^{-27} \text{ kg}) = 93.75 \times 10^{-27} \text{ kg}$$

Find the mass defect by subtracting the mass of the Fe nucleus from the constituent parts:

$$\Delta m = m_{\text{nucleons}} - m_{\text{Fe}}$$

$$\Delta m = 93.75 \times 10^{-27} \text{ kg} - 92.86 \times 10^{-27} \text{ kg} = 0.89 \times 10^{-27} \text{ kg}$$

To find the binding energy, we use Einstein's famous equation:

$$E = \Delta mc^2 = (0.89 \times 10^{-27} \text{ kg})(3 \times 10^8 \text{ m/s}^2) = 8.0 \times 10^{-11} \text{ J}$$

478. Conservation of charge: This is the same as conservation of atomic number or Z-number.

Conservation of nucleon number: This is the same as conservation of atomic mass number or A-number.

Conservation of mass/energy: This is equivalent to conservation of energy, realizing that mass is equal to the rest energy: $E = mc^2$.

Conservation of momentum: This is just like momentum collisions in mechanics. For example, during alpha decay, the unstable nucleus will shoot out an alpha particle at high speed, while the nucleus will recoil at slow speed due to its larger mass. Note that we can use the atomic mass number (A-number) as the mass of the reactants.

479. Alpha decay: A $_2^4\text{He}$ nucleus is ejected from the unstable nucleus, causing a loss of two protons and two neutrons for the remaining nucleus. The alpha particle leaves with kinetic energy; therefore, the mass of the products are less than that of the original nucleus. The nucleus will recoil in the opposite direction of that of the departing alpha particle.

Beta decay: An $_{-1}^{0}e$ electron is ejected from the unstable nucleus, causing a transformation of a neutron into a proton. The beta particle leaves with kinetic energy; therefore, the mass of the products are less than that of the original nucleus. The original nucleus will recoil in the opposite direction of that of the departing beta particle. This recoil will be less than that caused by an alpha particle due to the smaller mass of the beta particle.

Gamma decay: An excited state of a nucleus ejects a gamma ray as it transitions to a lower energy state. Gamma rays are massless and changeless and do not affect the nucleon number, or charge of the nucleus. The gamma ray takes energy away from the nucleus. Therefore, the mass of the nucleus decreases. In theory, the nucleus will recoil when the gamma ray departs, but the recoil will be very small.

480. Nuclear decay occurs when a nucleus is unstable because there is not enough binding energy to hold it together due to an excess number of nucleons. Mass is lost in the process (the mass defect increases), and the products of decay are more stable than the reactants.

481. Mass is lost in the process of decay. The mass defect increases during decay. This makes the products of decay more stable than the original nuclei.

482. Fission: Sometimes when a large nucleus is bombarded by a neutron, the nucleus becomes unstable and splits into two smaller nuclei and multiple neutrons. Both the nuclear charge and the number of nucleons are conserved in the process. The final products are more stable as mass is converted into binding energy. Some of the nuclear mass is also converted into kinetic energy of the reactants. The expelled neutrons can fission more nuclei, causing a cascading reaction (nuclear reactor and nuclear bomb).

Fusion: Given enough initial velocity, two small nuclei can be forced together past electrostatic repulsion to where the nuclear strong force will fuse them together to create a larger nucleus. During the process, all four conservations are obeyed. The final nucleus has less mass, as energy is released like in the sun.

483. **(A)** Not possible! The number of nucleons is not conserved.

(B) Not possible! Conservation of charge is violated.

484. **(A)** Fusion: $2_1^1\text{H} + 2_0^1\text{n} \rightarrow {}_2^4\text{He} + \text{energy}$

(B) Beta decay: $_{55}^{137}\text{Cs} \rightarrow {}_{56}^{137}\text{Ba} + {}_{-1}^{0}e$

(C) Alpha decay: $_{92}^{238}\text{U} \rightarrow {}_{90}^{234}\text{Th} + {}_2^4\alpha + \text{energy}$

(D) Fission: $^{1}_{0}n + ^{235}_{92}U \rightarrow ^{144}_{54}Xe + ^{90}_{38}Sr + 2^{1}_{0}n$

(E) Gamma decay: $^{12}_{6}C \rightarrow ^{12}_{6}C + \gamma$

485. Sample #1: approximately 3.5 years. Sample #2: approximately 2.5 minutes.

486. Half-life is the time it takes for one half of a radioactive sample to decay into a new element. After two half-lives, a quarter of the original radioactive substance will remain. This means that three quarters of the original material has transmuted into a new element.

487. 1.6 decays/min/g is approximately one-eighth of the original activity: (12.6/2/2/2 = 1.57). This is three half-lives. Therefore, the wood is approximately 17,000 years old: 5,730 years × 3 half-lives = 17,190 years.

488. **(A)** To find the energy released, first find the mass defect:

$$\Delta m = m_{Rn} - m_{Po} - m_{\alpha}$$
$$\Delta m = 222.018 \text{ u} - 218.009 \text{ u} - 4.002 \text{ u} = 0.007 \text{ u}$$

Then calculate the energy released:

$$E = \Delta mc^2 = (0.007 \text{ u})(1.66 \times 10^{-27} \text{ kg/u})(3 \times 10^8 \text{ m/s})^2 = 1.05 \times 10^{-12} \text{ J}$$

(B) Calculate the velocity of the alpha particle:

$$K = \frac{1}{2}mv^2$$

$$v = \sqrt{\frac{2K}{m}} = \sqrt{\frac{2(1.05 \times 10^{-12} \text{ J})}{(4.002 \text{ u})(1.66 \times 10^{-27} \text{ kg/u})}} = 1.78 \times 10^7 \text{ m/s}$$

Use conservation of momentum to find the velocity of the polonium nucleus. Since the original velocity of the radon atom is not given, we assume that it was stationary:

$$(p_{Rn})_{initial} = 0 = (p_{Po} + p_{\alpha})_{final}$$
$$p_{Po} = p_{\alpha}$$
$$mv_{Po} = mv_{\alpha}$$
$$(218.009 \text{ u})v_{Po} = (4.002 \text{ u})(1.78 \times 10^7 \text{ m/s})$$
$$v_{Po} = 3.27 \times 10^5 \text{ m/s}$$

(C) The half-life of radon is 3.8 days. Dividing 19 days by the 3.8 days half-life, we see that five half-lives have passed.

$$(m_{Rn})_{remaining} = \frac{(m_{Rn})_{original}}{2^{(\text{number of half-lives that have passed})}}$$

$$(m_{Rn})_{remaining} = \frac{20 \text{ kg}}{2^5} = \frac{20 \text{ kg}}{32} = 0.625 \text{ kg}$$

(D) Polonium is also radioactive with a 3.10-minute half-life. This is a much shorter half-life than that of radon. Most of the polonium will decay into a new element shortly after being created. So we can't really answer this question except to say that most of the radon that has decayed into polonium will already have been transmuted into some new element.

AP-Style Multiple-Choice Questions

489. (C) Special relativity tells us that when two observers are moving relative to each other, they will not necessarily agree on length and time. This becomes evident when we get up near the speed of light. We start to notice the effect around 0.1 c and faster. The only constant that all observers will agree on is the speed of light.

490. (D) The difference in radius is small. The radius of curvature for the ions is given by $r = \frac{mv}{qB}$. The radius is directly proportional to the mass and inversely proportional to the charge of the ions: $r \propto \frac{m}{q}$. A different isotope of neon would have a small difference in mass that could account for the small difference in radius. It is possible that the neon could have been singly, doubly, or triply ionized. However, this would change the charge by a whole-number factor. This would cause the different radii to be whole-number multiples of each other. Neon cannot have different numbers of protons without becoming a new element. This is not an interference pattern, so it cannot be formed by wave interference.

491. (C) The particle is most likely to be found at the highest positive/negative amplitude location of Ψ as a function of x. The particle will not be found at locations −15 nm and 0.0 nm.

492. (C) $^{237}_{93}\text{N} \rightarrow ^{205}_{81}\text{Tl} + 4\,^{0}_{-1}\beta + ^{A}_{Z}\alpha$. Solve for A and Z. A = 32 and Z = 16. Knowing that alpha particles have two protons and four nucleons, we can divide 32 by 4 or divide 16 by 2 to find out that we need 8 alpha particles to balance the equation.

493. (B and C) Conservation of charge is satisfied because the net charge before and after is zero. Conservation of momentum tells us that the initial momentum of the gamma ray must be equal to the net momentum of the two particles. The gamma ray has x-direction momentum and no y-direction momentum. This appears to be satisfied in the figure. The answer choice D is incorrect; we have to use conservation of mass/energy because most of the gamma ray energy has been converted into mass, not just kinetic energy of the two particles.

494. (C) Adding the excitation energy of 100 eV – 115 eV to the ground state energy, we get the electron energy range of (–22.4 eV) – (–7.4 eV). This will take the electrons of the gas from the ground state to the $n = 3$ and 4 energy levels. The highest frequency photon occurs when the electron jumps down in energy by the largest amount: $E = hf$. Therefore, the transition from $n = 3$ to $n = 2$ is the answer. Answer choices A and D are not possible because the electrons were not excited to the $n = 5$ level.

495. (A and C) A higher work function would account for lower energy electron emissions from A: $K_{max} = hf - \Phi$. Placing the plate closer to the light source would increase the number of photons captured by the plate, but it will not increase the energy of the photons striking the plate. Thus, it would produce more electrons, but the energy would remain the same for each electron.

496. (B and C) The discreet energy levels of electrons in an atom can be accounted for by the wave nature of the electron and constructive and destructive interference. The diffraction is a wave property. The photoelectric effect demonstrates the particle nature of light.

497. (D) The electrostatic repulsion of the protons is much stronger than the gravitational attraction of the nucleons. Thus, we need a stronger force to hold the nucleus together. Scientists call this the nuclear strong force.

498. (B) The energy of the gamma ray comes from the mass that is lost in the reaction. Adding the mass of the reactants, subtracting the mass of the products, and converting into energy via $E = mc^2$ gives us the mass defect that was converted into the energy of the gamma ray. This is an application of conservation of mass/energy.

AP-Style Free-Response Questions

499. (a) Helium has two protons and four neutrons. Completing the nuclear equation with conservation of nucleon number and conservation of charge, we get $^{5}_{3}\text{Li} \rightarrow {}^{4}_{2}\text{He} + {}^{1}_{1}\text{p} + 2.475 \text{ MeV energy}$. Therefore, lithium has three protons and two neutrons.

(b) Gravity pulls the nucleons together, but it is very weak. The protons are repelled by the electric force. The nuclear strong force is stronger than all of these forces and attracts all the nucleons together, making the nucleus stable.

(c) The kinetic energy comes from the mass that is lost in the reaction that is converted into energy:

$$m_{Li}c^2 = m_{He}c^2 + m_p c^2 + 2.475 \text{ MeV}$$

(d) Conservation of momentum applies. The lithium was at rest. Therefore, the helium and proton must move off in opposite directions. The helium has more mass and will move more slowly than the lighter proton.

$$p_{initial} = p_{final}$$
$$0 = mv_{He} + mv_p$$

(e) First convert to standard units:

Proton: $(1.0073 \text{ u})(1.66 \times 10^{-27} \text{ kg/u}) = 1.672 \times 10^{-27}$ kg
Helium: 4.0012 u $= 6.642 \times 10^{-27}$ kg.
Converting energy to joules: $(2.475 \times 10^6 \text{ eV})(1.6 \times 10^{-19} \text{ J/eV}) = 3.96 \times 10^{-13}$ J

Use conservation of mass/energy:

$$m_{Li}c^2 = m_{He}c^2 + m_p c^2 + 2.475 \text{ MeV}$$

$$m_{Li} = m_{He} + m_p + \frac{2.475 \text{ MeV}}{c^2}$$

Insert the values into the equation, and use 3.0×10^8 m/s for the speed of light:

Lithium mass $= 8.318 \times 10^{-27}$ kg

Note that only 0.004×10^{-27} kg of mass is needed to create the 2.475 MeV of energy in this reaction.

(f) They will need to manufacture it. The half-life is very short, and this isotope will decay away into other atoms in less than a fraction of a second.

500. (a) First convert the wavelengths of light into energy in electron volts:

$$E = hf = \frac{hc}{\lambda}$$

248 nm $= 5$ eV, 400 nm $= 3.1$ eV, 650 nm $= 1.9$ eV

With this information, we can build the energy level diagram. Remember that absorbing photons causes the electron to jump upward. Starting in the ground state of –8 eV and adding a 5-eV absorbed photon, we see that there is an energy level at –3 eV. To produce the 400-nm and 600-nm light, there must be an intermediate level between the ground state and the –3-eV level. There are two acceptable ways to draw this intermediate energy level, as shown in the two figures, depending on whether the 650-nm photon is emitted first or second. Be sure to label the energies of each level.

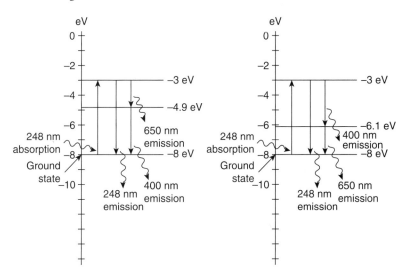

(b) No, it will not eject electrons. The 650-nm photon has even less energy than the 400-nm photon and will be below the threshold needed to eject and electron:

$$E = hf = \frac{hc}{\lambda}$$

(c) i. We need to convert electron energy into joules of kinetic energy:

$$(2.71 \text{ eV})(1.6 \times 10^{-19} \text{ J/eV}) = 4.34 \times 10^{-19} \text{ J}$$

Calculate the velocity of the electron. Remember that the mass of an electron is 9.11×10^{-31} kg:

$$K = \frac{1}{2}mv^2$$

$$v = \sqrt{\frac{2K}{m}} = 9.76 \times 10^5 \text{ m/s}$$

Calculate the de Broglie wavelength. Remember that $h = 6.63 \times 10^{-34}$ J·s:

$$\lambda = \frac{h}{p} = \frac{h}{mv} = 7.46 \times 10^{-10} \text{ m}$$

ii. The equation $\sin\theta = \frac{m\lambda}{d}$ shows us that the ratio $\frac{\lambda}{d}$ needs to be less than 1 for an interference pattern to form. Any larger ratio will cause the angle θ to be larger than 90 degrees, and no pattern will form. This means the wavelength must be smaller than the spacing, but not too small or the angle will be so small that the pattern will be too small to see. The electrons will form an interference pattern, as the waves are on the same order of magnitude and smaller than the opening spacing. As a side note, none of the light emitted from the gas will form an interference pattern because the wavelengths are much larger than the opening spacing.